Lecture Notes in Computer Science   6405

Commenced Publication in 1973
Founding and Former Series Editors:
Gerhard Goos, Juris Hartmanis, and Jan van Leeuwen

Kedar Namjoshi   Andreas Zeller   Avi Ziv (Eds.)

Hardware and Software: Verification and Testing

5th International Haifa Verification Conference, HVC 2009
Haifa, Israel, October 19-22, 2009
Revised Selected Papers

 Springer

Volume Editors

Kedar Namjoshi
Bell Laboratories, Alcatel-Lucent
600-700 Mountain Avenue, Murray Hill, NJ 07974, USA
E-mail: kedar@research.bell-labs.com

Andreas Zeller
Saarland University
Campus E1, 66123 Saarbrücken, Germany
E-mail: zeller@cs.uni-saarland.de

Avi Ziv
IBM Research Laboratory
Mount Carmel, Haifa 31905, Israel
E-mail: aziv@il.ibm.com

ISSN 0302-9743   ISSN 1611-3349
ISBN 978-3-642-19236-4   ISBN 978-3-642-19237-1 (eBook)
DOI 10.1007/978-3-642-19237-1
Springer Heidelberg Dordrecht London New York

Library of Congress Control Number: 2011920830

CR Subject Classification (1998): D.2.4-5, D.2, D.3, F.3

LNCS Sublibrary: SL 2 – Programming and Software Engineering

Typesetting: Camera-ready by author, data conversion by Scientific Publishing Services, Chennai, India

Printed on acid-free paper

Springer is part of Springer Science+Business Media (www.springer.com)

Preface

This volume holds the proceedings of HVC 2009. The Haifa Verification Conference is unique in bringing together research communities from formal and dynamic verification of hardware and software systems. It thus encourages both the recognition of common core questions and a healthy exchange of ideas and methods across domains. The attendees at HVC come from academia, industrial research labs, and industry, resulting in a broad range of perspectives.

The program for this year was chosen from 23 submissions. While we faced an unexpected drop in submissions, the resulting program was of a high quality. The paper by Anna Moss and Boris Gutkovich on "Functional Test Generation with Distribution Constraints" was chosen for the Best Paper Award. The HVC Award, given to the most promising contribution in the fields of software and hardware verification and test in the past five years, was given to Patrice Godefroid, Nils Klarlund, and Koushik Sen for their work on "DART: Directed Automated Random Testing."

The program included an outstanding set of keynote and invited talks. David Harel from the Weizmann Institute of Science spoke on "Can We Verify an Elephant?"; Mark Harman from CREST centre at King's College London, spoke on "The SBSE Approach to Automated Optimization of Verification and Testing;" and Harry Foster from Mentor Graphics, spoke on "Pain, Possibilities, and Prescriptions Industry Trends in Advanced Functional Verification." Tutorials were organized on "Post-Silicon Validation and Debugging," with Amir Nahir and Allon Adir (IBM), Rand Grey and Shmuel Branski (Intel), and Brad Quinton (University of British Columbia); "Satisfiability Modulo Theories" with Ofer Strichman (Technion); and "Constraint Satisfaction" with Eyal Bin (IBM). We would like to thank the speakers for putting together interesting and informative talks.

The conference was held at IBM's Research Labs at Haifa. We would like to thank the many people who were involved; in particular, Vered Aharon, who made sure that the conference ran smoothly each day. The Program Committee worked hard to put together the conference program; we thank them for their efforts. The HVC Organizing Committee provided considerable help and perspective. The HVC Award Committee, which was chaired by Sharad Malik (Princeton) and included Holger Hermanns (Saarland), Sarfraz Khurshid (University of Texas, Austin), Natarajan Shankar (SRI), and Helmut Veith (TU Darmstadt), did a wonderful job in picking a particularly deserving paper for the award among the many good candidates. Finally, we would like to thank all those who participated in the conference and made it an exciting and enjoyable event.

December 2010

Avi Ziv
Kedar Namjoshi
Andreas Zeller

Conference Organization

General Chair

Avi Ziv   IBM Research

Program Chairs

Kedar Namjoshi   Bell Labs, Alcatel-Lucent
Andreas Zeller   Saarland University

Program Committee

Eyal Bin	IBM Research, Israel
Roderick Bloem	TU Graz, Austria
Hana Chockler	IBM Research, Israel
Myra Cohen	University of Nebraska-Lincoln, USA
Christoph Csallner	University of Texas at Arlington, USA
Kerstin Eder	Bristol University, UK
Steven German	IBM Research, USA
Patrice Godefroid	Microsoft, USA
Orna Grumberg	Technion, Israel
Shankar Hemmady	Synopsys, USA
Gerard Holzmann	NASA, USA
Vineet Kahlon	NEC, USA
Sharon Keidar-Barner	IBM Research, Israel
Orna Kupferman	Hebrew University, Israel
Doron Peled	Bar Ilan University, Israel
Andreas Podelski	University of Freiburg, Germany
Gil Shurek	IBM Research, Israel
Scott Stoller	Stony Brook University, USA
Shmuel Ur	IBM Research, Israel
Tao Xie	North Carolina State University, USA
Eran Yahav	IBM Research, USA
Karen Yorav	IBM Research, Israel

HVC Award Committee

Sharad Malik	Princeton University, USA (Chair)
Holger Hermanns	Saarland University, Germany
Sarfraz Khurshid	University of Texas, Austin, USA
Natarajan Shankar	SRI International, USA
Helmut Veith	Technische Universität Darmstadt, Germany

Local Organization

Vered Aharon   IBM Haifa

Sponsors

The Organizing Committee gratefully acknowledges the support provided by IBM Haifa Research Labs and Cadence Israel.

External Reviewers

Allon Adir	Amir Nahir	Ariel Cohen
Eitan Marcus	Hana Chockler	Ilan Beer
Ishtiaque Hussain	Jianjun Zhao	Karine Even
Madan Musuvathi	Michael Case	Michael Gorbovitski
Michal Rimon	Mithun Acharya	Nir Piterman
Ronny Morad	Sitvanit Ruah	Stefan Schwoon
Viresh Paruthi	Wujie Zheng	Yael Meller
Yakir Vizel	Yoav Katz	

Table of Contents

I   Keynote and Invited Talks

II   Research Papers

Can We Verify an Elephant?

David Harel

Weizmann Institute of Science

Abstract. The talk shows the way techniques from computer science and software engineering can be applied beneficially to research in the life sciences. We will discuss the idea of comprehensive and realistic modeling of biological systems, where we try to understand and analyze an entire system in detail, utilizing in the modeling effort all that is known about it. I will address the motivation for such modeling and the philosophy underlying the techniques for carrying it out, as well as the crucial "verification" question of when such models are to be deemed valid, or complete. The examples I will present will be from among the biological modeling efforts my group has been involved in: T cell development in the thymus, lymph node behavior, organogenesis of the pancreas, fate determination in the reproductive system of C. elegans, and a generic cell model. The ultimate long-term "grand challenge" is to produce an interactive, dynamic, computerized model of an entire multi-cellular organism, such as the C. elegans nematode worm, which is complex, but well-defined in terms of anatomy and genetics. The challenge is to construct a full, true-to-all-known-facts, 4-dimensional, fully animated model of the development and behavior of this worm (or of a comparable multi-cellular animal), which is easily extendable as new biological facts are discovered.

K. Namjoshi, A. Zeller, and A. Ziv (Eds.): HVC 2009, LNCS 6405, p. 1, 2011.
© Springer-Verlag Berlin Heidelberg 2011

Pain, Possibilities, and Prescriptions Industry Trends in Advanced Functional Verification

Harry Foster

Mentor Graphics

Abstract. Today, many forces at play contribute to the gap between what we can fabricate (silicon capacity) and what we have time to design. In addition, there are forces at play that contribute to a gap between what we can design and what we realistically have time to verify (within a project's schedule). Nonetheless, we tape out complex systems all the time. Hence, the question arises, is the productivity gap real? And if so, what can we do to minimize its effects? This talk provides a statistical analysis of today's industry trends in the adoption of advanced functional verification (AFV) techniques, and then offers new models for improving AFV maturity within an organization.

K. Namjoshi, A. Zeller, and A. Ziv (Eds.): HVC 2009, LNCS 6405, p. 2, 2011.

The SBSE Approach to Automated Optimization of Verification and Testing

Mark Harman

CREST Centre at King's College London

Abstract. The aim of Search Based Software Engineering (SBSE) research is to provide automated optimization for activities right across the Software Engineering spectrum using a variety of techniques from the metaheuristic search, operations research and evolutionary computation paradigms. The SBSE approach has recently generated a great deal of interest, particularly in the field of Software Testing. There is a natural translation from test input spaces to the search spaces on which SBSE operates and from test criteria and goals to the formulation of the fitness functions with which the search based optimization algorithms are guided. This makes SBSE compelling and generic; many testing problems can be formulated as SBSE problems. This talk will give an overview of SBSE applications to testing, verification and debugging with some recent results and pointers to open problems and challenges.

K. Namjoshi, A. Zeller, and A. Ziv (Eds.): HVC 2009, LNCS 6405, p. 3, 2011.
© Springer-Verlag Berlin Heidelberg 2011

DART: Directed Automated Random Testing

Koushik Sen

UC Berkeley

Abstract. Testing with manually generated test cases is the primary technique used in industry to improve reliability of software–in fact, such testing is reported to account for over half of the typical cost of software development. I will describe directed automated random testing (also known as concolic testing), an efficient approach which combines random and symbolic testing. Concolic testing enables automatic and systematic testing of programs, avoids redundant test cases and does not generate false warnings. Experiments on real-world software show that concolic testing can be used to effectively catch generic errors such as assertion violations, memory leaks, uncaught exceptions, and segmentation faults. From our initial experience with concolic testing we have learned that a primary challenge in scaling concolic testing to larger programs is the combinatorial explosion of the path space. It is likely that sophisticated strategies for searching this path space are needed to generate inputs that effectively test large programs (by, e.g., achieving significant branch coverage). I will present several such heuristic search strategies, including a novel strategy guided by the control flow graph of the program under test.

K. Namjoshi, A. Zeller, and A. Ziv (Eds.): HVC 2009, LNCS 6405, p. 4, 2011.

Reduction of Interrupt Handler Executions for Model Checking Embedded Software

Bastian Schlich[1], Thomas Noll[2], Jörg Brauer[1], and Lucas Brutschy[1]

[1] Embedded Software Laboratory, RWTH Aachen University
Ahornstr. 55, 52074 Aachen, Germany
[2] Software Modeling and Verification Group, RWTH Aachen University
Ahornstr. 55, 52074 Aachen, Germany

Abstract. Interrupts play an important role in embedded software. Unfortunately, they aggravate the state-explosion problem that model checking is suffering from. Therefore, we propose a new abstraction technique based on partial order reduction that minimizes the number of locations where interrupt handlers need to be executed during model checking. This significantly reduces state spaces while the validity of the verification results is preserved. The paper details the underlying static analysis which is employed to annotate the programs before verification. Moreover, it introduces a formal model which is used to prove that the presented abstraction technique preserves the validity of the branching-time logic CTL*-X by establishing a stutter bisimulation equivalence between the abstract and the concrete transition system. Finally, the effectiveness of this abstraction is demonstrated in a case study.

1 Introduction

Embedded systems frequently occur as part of safety-critical systems. Full testing of these systems is often not possible due to fast time to market, uncertain environments, and the complexity of the systems. Model checking has been recognized as a promising tool for the analysis of such systems. A major problem for the application of model checking is the state explosion. When model checking embedded-systems software, interrupts are a major challenge. They are important as many features of embedded systems are implemented using interrupts, but they have a considerable impact on the size of the state space. Whenever they are enabled, they can interact with the main program and influence the behavior of the overall system.

To make model checking applicable to embedded systems software, we developed a model checker for microcontroller assembly code called [MC]SQUARE [1]. This model checker works directly on the assembly code of the program and automatically applies abstraction techniques such as delayed nondeterminism [2] and delayed nondeterminism for interrupts [3] to tackle the state-explosion problem. This paper describes a new abstraction technique called *interrupt handler execution reduction* (IHER), which is based on the idea of partial order reduction (POR). It reduces the number of program locations at which the possible

K. Namjoshi, A. Zeller, and A. Ziv (Eds.): HVC 2009, LNCS 6405, pp. 5–20, 2011.

execution of interrupt handlers (IHs) has to be considered. This can greatly reduce state spaces built during model checking.

The idea behind IHER is similar to the one behind POR (cf. Sect. 6), but the algorithms used are different due to the fact that the pseudo parallelism introduced by IHs significantly differs from concurrent threads in its asymmetry. Threads can block other threads and control can nondeterministically change between threads. IHs, however, can only interrupt the main program, but they cannot be interrupted by the main program. The instructions of the main program have to be executed whereas the execution of IHs is usually nondeterministic. Moreover in [MC]SQUARE, IHs are required to be executed atomically because if IHs can mutually be interrupted, a stack collision will eventually occur as the stack that stores the return addresses is bounded on real microcontrollers. Hence, we model IHs as atomic actions in IHER. Consequently, as IHs do not necessarily terminate due to loops or usage of microcontroller features, we cannot guarantee termination of atomic actions. In contrast, in POR it is assumed that atomic actions always terminate.

The contribution of this paper is twofold. We have developed a static analysis framework for microcontroller assembly code that forms the basis for IHER. Furthermore, we have developed a dynamic part that applies IHER during model checking. The static analysis identifies program locations at which the execution of IHs can be prevented because they do not influence the (visible) behavior of the system or software, respectively. During model checking, the execution of IHs is blocked at these locations. As we will see, this abstraction technique guarantees a stuttering bisimulation equivalence between the concrete and the abstract transition system. Therefore, it preserves the validity of CTL*-X [4] formulas.

The paper is structured as follows. First, [MC]SQUARE is introduced in Sect. 2. Then, Sect. 3 explains the general idea of our abstraction technique and details the applied algorithms. Section 4 presents a formal model and gives a sketch of the proof that the abstraction technique presented in this paper actually preserves a stuttering bisimulation equivalence. The effectiveness of the technique is demonstrated in the case study described in Sect. 5. Related work, particularly with respect to POR, is presented in Sect. 6.

2 [MC]SQUARE

[MC]SQUARE [1] is a model checker for microcontroller assembly code. It can verify code for five different microcontrollers, namely ATMEL ATmega16 and ATmega128, Infineon XC167, Intel MCS-51, and Renesas R8C/23. It accepts programs given in different binary-code file formats such as ELF or Intel Hex Format and, additionally, it reads the corresponding C code if it is available. [MC]SQUARE processes specifications given in CTL [4], which can include propositions about general purpose registers, I/O registers, general memory locations, and the program counter. (Depending on the applied abstraction techniques, propositions about the program counter may be disallowed.) If debug information is available, specifications can also include propositions about C variables.

[MC]SQUARE uses explicit model checking algorithms, but the states are partly symbolic. That is, they do not represent single concrete states but sets of concrete states, and are introduced by abstractions of the microcontroller memory. In [MC]SQUARE, we have modeled different abstractions of the memory that vary with respect to the degree of abstraction. Beside these memory-oriented methods, we have also implemented several general purpose abstraction techniques such as dead variable reduction and path reduction [5]. It is important to notice that [MC]SQUARE always creates an over-approximation of the behavior shown by the real microcontroller. Depending on the applied abstraction techniques, [MC]-SQUARE preserves the validity of CTL, the universal fragment of CTL (ACTL) [4], or ACTL-X, which refers to ACTL without the *next* operator.

Figure 1 shows the model checking process that is applied by [MC]SQUARE. First, the binary code, the C code (if available), and the formula are parsed and transformed into their internal representations. Then, the static analyzer is executed and the program is annotated using information from the assembly code, the debug information, and the CTL formula. These annotations are later used by the simulator to reduce the state space.

The static analyzer performs several analyses as described by Schlich [1]. A major challenge for the analysis of assembly code are indirect references to the memory. As most of these are caused by stack-handling operations, a stack analysis is employed to determine the values of the stack pointer in order to restrict the memory regions that can be accessed [6]. Other indirect references are rarely used. To generate an over-approximation, we assume in these cases that indirect references can access the complete memory of the microcontroller.

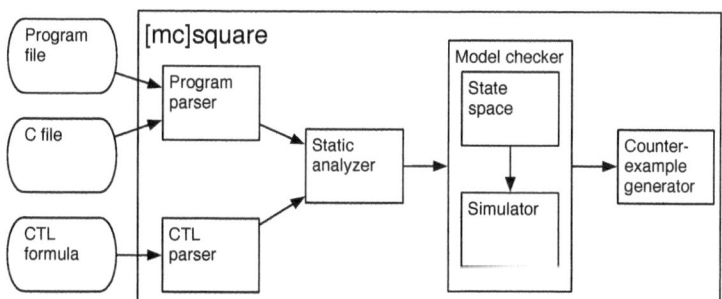

Fig. 1. Model checking process applied in [MC]SQUARE

In the next step, [MC]SQUARE performs model checking. Currently, we have implemented two different algorithms: one for checking invariants, and one on-the-fly CTL model checking algorithm described by Heljanko [7]. The model checker requests states from the state space. If successors of a state are not yet created, the state space uses the simulator to generate them on-the-fly.

The simulator natively handles nondeterminism and creates an over-approximation of the behavior shown by the real microcontroller. Within the simulator component, we have modeled the different microcontrollers. Creation of successors is done by means of interpretation. A state is loaded into the model of the microcontroller, and then all its possible successors are generated. A state can have more than one successor because interrupts can occur while executing the program and because input can be read from the environment or from devices with nondeterministic behavior such as timers.

If the model checker refutes the property under consideration, the counterexample generator creates a counterexample, which is also optimized. That is, loops and other unneeded parts are removed to ease its comprehension. The counterexample is presented in the assembly code, in the C code (if available), in the control flow graph of the assembly code, and as a state space graph.

3   Reduction of Interrupt Handler Executions

The execution of IHs has a significant impact on state space sizes when model checking microcontroller programs. Interrupts introduce pseudo parallelism in microcontroller programs as they can possibly occur at every program location where they are enabled (cf. Sect. 1). Thus, state spaces can grow exponentially with the number of interrupts used. Similarly to the observation that led to the partial order reduction technique (cf. Sect. 6), we observed that the execution of IHs does not always influence the behavior of a program. In the following, an abstraction technique is described that reduces the number of locations where IHs have to be considered. First, the general idea is presented, then, details of the applied analysis are given, and in the end, the application of this technique is demonstrated using an example.

3.1   General Idea

We have developed an abstraction technique called *interrupt handler execution reduction*, which reduces the number of IH executions by blocking IHs at program locations where there is no dependency between certain IHs and the program. There is a dependency if either one influences the other or the visible behavior of the program is changed. An IH influences a program location if it, for example, writes a memory location that is accessed by the program location. Here, an access refers to both a reading or writing reference to a memory location. On the other hand, a program location influences an IH if it, for instance, enables or disables interrupts. The visible behavior of the program is changed by a visible action if a memory location is written that is used in an atomic proposition (AP). The same applies for dependencies between IHs.

When using this abstraction technique, propositions about the program counter are not allowed because the program counter is changed at all program locations, and therefore, IHs could never be suppressed. In the analysis, the execution of IHs is assumed to be atomic, and therefore, IHs are treated as single instructions. Our idea, however, can easily be extended to the case that IHs are interruptible by treating each IH the same way as the main process is treated.

As IHs can possibly contain divergent loops, termination of IHs cannot be guaranteed. To preserve the validity of specifications with respect to our abstraction technique, divergent behavior has to be observable both in the concrete and the abstract model (see Sect. 4). Hence, IHs have to be executed at least once between two visible actions.

We additionally require that an interrupt can occur arbitrarily often at a single program location because at this location it has to mimic all possible behaviors to create an over-approximation of the real behavior. On the real hardware, for some microcontrollers such as the ATMEL ATmega16, the execution of an IH is always followed by the execution of an instruction of the program. Allowing an arbitrary number of occurrences adds additional behavior and thus leads to an over-approximation, but it again reduces state spaces.

The IHER technique comprises two parts: a static analysis that annotates the program, and a dynamic part that uses the annotations during model checking to suppress the execution of IHs where they do not need to be considered. The next section details the static analysis and the last section provides an example.

3.2   Static Analysis

As a prerequisite for determining program locations where IHs can be blocked, [MC]SQUARE employs a sequence of different context-sensitive static analyses and combines their results as detailed by Schlich [1]. First, the control flow graph (CFG) of the program is built and all program locations are annotated with the sets of live variables, reaching definitions, and the status of interrupt registers, that is, the information whether certain interrupts are enabled or disabled. During these analyses, information about the stack is used to limit the over-approximation. As their results potentially influence each other, these analyses are conducted within a loop until a fixed point is reached. Using the information that was obtained in this way, the analysis for the IHER abstraction technique is applied. It consists of the following four steps:

1. Detect dependencies between IHs
2. Detect dependencies between program locations and IHs
3. Refine results
4. Label blocking locations

In the following, these four steps are detailed.

Detect Dependencies between IHs. In the first step of the analysis, dependencies between IHs are identified. This is formalized by the relation $\bowtie \subseteq IH \times IH$ where i, $j \in IH$ *depend on each other*, denoted $i \bowtie j$, if one of the following conditions holds:

- one enables or disables the other,
- one writes a memory location accessed by the other, or
- one writes a memory location used in an AP.

This relation is obviously symmetric. If one IH enables or disables another IH, all possible interleavings between both are relevant. Therefore, not only the enabled/disabled IH has to be executed if the enabling/disabling IH is executed but also vice versa as otherwise behavior could get lost. This also applies if one IH writes a memory location that is accessed by another IH. Note that in the last condition only one IH is mentioned. Thus, if there is one IH that writes a memory location that is used in an AP, all IHs depend on each other. An IH that writes an AP is related to all IHs including those that do not write APs because its execution could be prevented by a non-terminating IH. This includes the case of two IHs that both write APs: they are related because they both alter the visible behavior of the program, and thus, all their possible interleavings are relevant.

The transitive-reflexive closure of $\bowtie$ is denoted by $\bowtie^*$ and induces a partitioning of IH. This partitioning is used in the following way. Whenever one of the IHs has to be executed, all other IHs in its equivalence class have to be executed as well. The algorithm to compute the dependency relation performs a nested iteration over all IHs based on the conditions described above.

Detect Dependencies between Program and IHs. In the second step of the analysis, [MC]SQUARE determines dependencies between the program and the IHs and identifies program locations where interrupts have to be executed. There exists a dependency between a program location and an IH if either one influences the other or the program behavior is visibly changed. The latter is the case if an instruction or an IH writes memory locations used in APs.

To detect the dependencies between the program and the IHs, [MC]SQUARE marks specific program locations with the following two labels: *execution* and *barrier*. The label *execution* implies that there exists a dependency between the preceding program location and an IH, and thus, this IH needs to be executed eventually. The label *barrier* denotes that there exists a dependency between that program location and an IH, and therefore, this IH needs to be executed before the instruction at that location is executed. Otherwise, visible behavior could get lost. In the later refinement step, label *execution* can be moved until a label *barrier* is reached.

Let program location k be a direct predecessor of program location l. Formally, for each $i \in$ IH, l is labeled with *execution$_i$* if one of the following conditions is satisfied:

- k enables or disables i,
- k writes a memory location that is accessed by i, or
- k writes a memory location that is used in an AP.

These conditions are similar to the conditions for dependencies between IHs. If k enables or disables an IH, this IH has to be executed eventually to exhibit the changed behavior. The same applies if k writes a memory location that is accessed by an IH. As interrupts are deactivated in the initial program location, a program location has to enable interrupts before they can influence the program or change the visible behavior of the program. Note that in the last condition only k is mentioned and not a specific IH. If k writes an AP, each IH has to be executed afterwards because the execution of an IH could either prevent the execution of another instruction that writes an AP or the IH could itself write an AP. If l is labeled with $execution_i$, it is also labeled with $execution_j$ $\forall j \in [i]_{\bowtie^*}$ because all IHs of the same equivalence class have to be executed at the same location.

For each $i \in$ IH, a program location l is labeled with $barrier_i$ if one of the following conditions holds:

- i writes a memory location that is accessed by l,
- l enables or disables i,
- l writes a memory location that is accessed by i, or
- l writes a memory location that is used in an atomic proposition.

The first condition is different from the conditions for label $execution_i$. If i writes a memory location that is accessed by l, i has to be executed before l is executed because otherwise a possibly changed behavior could get lost: the execution of i after the execution of l could no longer influence the execution of l. This condition shows the asymmetry between the program and the IHs. The remaining conditions are duals of the conditions for label $execution_i$. They are used to guarantee that a possibly changed behavior is finally considered. If l is labeled with $barrier_i$, it is also labeled with $barrier_j$ $\forall j \in [i]_{\bowtie^*}$.

Refine Results. In the refinement step, [MC]SQUARE tries to reduce state spaces further by moving $execution_i$ labels until their execution is actually required. This is possible because in the previous step, [MC]SQUARE only locally labeled program locations where IH behavior was changed, but did not check whether their changed behavior actually influences the program. During refinement, the context is taken into account. An IH and all dependent IHs do not have to be executed if all behavior relevant to the specification and the program is created through their execution at another program location. Therefore, it is sufficient to execute IHs at only one of these locations. In the refinement step, [MC]SQUARE moves labels $execution_i$ forward until one of the following conditions holds:

- a program location labeled with $barrier_i$ is reached,
- a loop entry is found, or
- a loop exit is found.

This further reduces state spaces by postponing the execution of IHs until required. The label $execution_i$ cannot be moved over a program location labeled with $barrier_i$ because it either influences the next instruction or the next instruction influences its behavior, and it has not yet been executed. Furthermore, a label $execution_i$ is not moved into a loop because this would possibly increase the size of the state space. Moreover, it is not moved out of a loop because loop termination cannot be guaranteed and divergent behavior has to be preserved in the abstract system.

Label Blocking Locations. In the last step, all program locations are labeled with IHs that can be blocked at the corresponding program location. An IH can be blocked at a program location if its execution is not required. Thus, a program location is labeled with $blocking_i$ if it is not labeled with $execution_i$.

3.3   An Example

To illustrate the IHER abstraction technique, we give an example. We apply this analysis to the program shown in Fig. 2(a) and the IH presented in Fig. 2(b). In the main program, interrupts are first enabled and then some calculations are executed on registers r1, r2, and r3. The IH accesses only register r1 and doubles its value. No atomic propositions are used in this example.

l_0	SEI	enable interrupts		i_0	ADD r1,r1	$r_1 \leftarrow r_1 + r_1$
l_1	LD r2,5	$r_2 \leftarrow 5$		i_1	RETI	return
l_2	ADD r1,r2	$r_1 \leftarrow r_1 + r_2$				
l_3	MOV r2,r1	$r_2 \leftarrow r_1$				
l_4	MOV r3,r2	$r_3 \leftarrow r_2$				
l_5	CLI	disable interrupts				
l_6	RJMP -1	self loop				
	(a) Main program				(b) Interrupt handler	

Fig. 2. Assembly code of the main program and the interrupt handler (excerpt)

In this example, only one IH is used, and therefore, the first step of the analysis can be omitted. In the second step, we label the program locations with *execution* and *barrier*. Here, we omit the indices for clarity. The resulting labeled CFG is depicted in Fig. 3(a). White circles represent program locations without labels, white octagons represent locations labeled with *barrier*, and grey nodes represent program locations labeled with *execution*. Edges are labeled with the corresponding instruction or IH respectively.

Locations l_1, l_3, and l_6 are labeled with *execution* because their preceding instructions influence the IH. Locations l_2 and l_3 are labeled with *barrier* as the IH influences the current instruction. Hence, the IH has to be executed not later than at these locations. Locations l_0 and l_5 are labeled with *barrier* because interrupts are enabled or disabled by the respective instruction.

In the refinement step, *execution* labels are moved forward until either a *barrier* label or a loop is reached. The result of this step for the program is shown in Fig. 3(b). Here, only the *execution* label of l_1 is moved to l_2 because l_2 is a barrier. The *execution* label in l_6 cannot be moved due to the self loop.

These labels are then translated into blocking locations. In Fig. 4 the differences in IH execution with and without IHER are shown. Figure 4(a) shows that without applying IHER, the IH is executed at five program locations because interrupts are disabled in l_0 and l_6. The application of IHER leads to the execution of the IH at only two locations as depicted in Fig. 4(b).

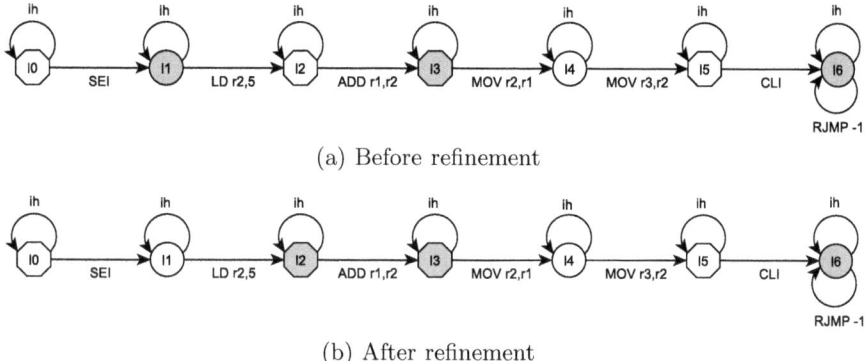

(a) Before refinement

(b) After refinement

Fig. 3. CFGs of the code shown in Fig. 2

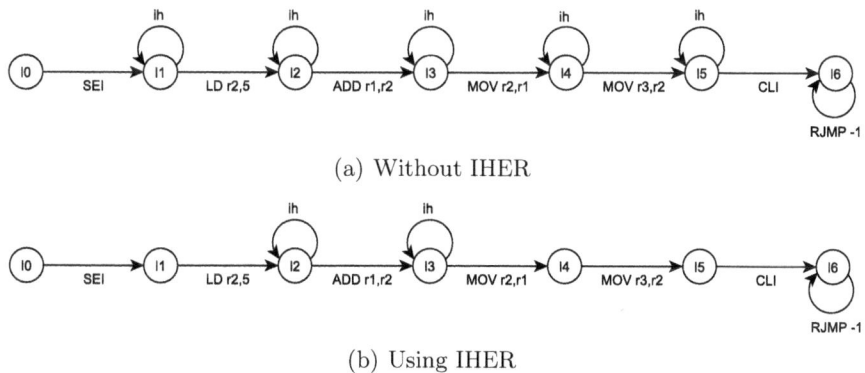

(a) Without IHER

(b) Using IHER

Fig. 4. Comparison of IH executions for program shown in Fig. 2

4   Formal Model and Correctness Proof

This section introduces the formal model on which the correctness proof of our abstraction technique is based. It is defined in two steps: (1) The syntactic structure of the microcontroller program is represented by its CFG, which consists of the program locations connected by control flow edges. Here, each edge carries an action label and a Boolean expression. The former represents the execution of either a single machine instruction or of a complete IH. The latter acts as a guard controlling the execution of, for instance, conditional branching instructions or IHs in dependence of the memory state. Note that single instructions of IHs are not considered as we assume their execution to be atomic. (2) Semantics is involved by associating with every action a mapping on the data space of the program, that is, the memory contents. This gives rise to a *labeled transition system* in which each state is given by a program location and a data state, where the latter represents the contents of general-purpose registers, I/O registers, and memory locations. Thus, the correctness proof boils down to showing that the original and the reduced CFG yield equivalent labeled transition systems.

4.1   The Formal Model

Formally, the CFG of the program is given by $G = (L, l_0, A, B, \longrightarrow)$ where

- L is a finite set of *program locations*,
- $l_0 \in L$ is the *initial location*,
- A is a finite set of *actions*,
- B is a finite set of *guards*, and
- $\longrightarrow \subseteq L \times A \times B \times L$ is the *control flow relation* (where each entry is represented as $l \xrightarrow{a,b} l'$ with $l, l' \in L$, $a \in A$, and $b \in B$).

The introduction of guards allows to model, e.g., conditional branching instructions by two transitions with the same action and different guards, which indicate the outcome of the evaluation of the condition.

The semantics of a CFG is determined by associating with every action $a \in A$ a mapping $[\![a]\!] : D \to 2^D$, and with every guard $b \in B$ a mapping $[\![b]\!] : D \to \mathbb{B}$. Here, D stands for the *data space*, that is, the finite set of *memory states* of the program. Interpreting $[\![a]\!](d)$ as a set of memory states allows us to model the non-deterministic nature of certain instructions, such as reading operations on input registers. Each of these sets is required to be non-empty and finite. A set is a singleton if the respective action is deterministic.

Applying this semantics to the given CFG G yields a *labeled transition system* $T(G) = (S, s_0, A, \Longrightarrow, P, \lambda)$, which is defined as follows:

- $S := L \times D$ is the finite set of *states*,
- $s_0 := (l_0, d_0) \in S$ is the *initial state* where $d_0 \in D$ stands for the initial data state,
- A is the finite set of *actions* (as before),
- $\Longrightarrow \subseteq S \times A \times S$ is the *transition relation*, given by: whenever $l \xrightarrow{a,b} l'$ in G and $d \in D$ such that $[\![b]\!](d) = \mathsf{true}$, then $(l, d) \xRightarrow{a} (l', d')$ for every $d' \in [\![a]\!](d)$,
- P is a finite set of *atomic propositions*, and
- $\lambda : S \to 2^P$ is the *property labeling*.

4.2   Correctness of the Abstraction

As explained in Sect. 3, IHER reduces the state space of the system by blocking IHs at program locations where they are independent of the main program. In other words, it removes certain transitions from the CFG (but keeps all locations), leading to a reduced graph $G_\sharp = (L, l_0, A, B, \longrightarrow_\sharp)$ with $\longrightarrow_\sharp \subseteq \longrightarrow$. According to the previous definition, $G_\sharp$ then yields a reduced labeled transition system $T(G_\sharp) = (S_\sharp, t_0, A, \Longrightarrow_\sharp, P, \lambda_\sharp)$ with $S_\sharp \subseteq S$, $t_0 = s_0$, $\Longrightarrow_\sharp \subseteq \Longrightarrow$, and $\lambda_\sharp = \lambda|_{S_\sharp}$.

We will now establish the correctness of our abstraction technique by showing that the original and the reduced transition system are equivalent. More concretely we will see that $T(G)$ and $T(G_\sharp)$ are related by a *divergence-sensitive stutter bisimulation*, implying that our abstraction mapping preserves the validity of formulas in CTL*-X [8].

We begin with the definition of a *stutter bisimulation* [9], which is a binary relation $\rho \subseteq S \times S_\sharp$ such that $s_0 \rho t_0$ and, for all $s\rho t$,

- $\lambda(s) = \lambda_\sharp(t)$,
- if $s \overset{a}{\Longrightarrow} s'$ with $(s', t) \notin \rho$, then there exists a path $t \overset{a_0}{\Longrightarrow}_\sharp u_1 \overset{a_1}{\Longrightarrow}_\sharp \ldots \overset{a_{n-1}}{\Longrightarrow}_\sharp u_n \overset{a_n}{\Longrightarrow}_\sharp t'$ with $n \geq 0$, $s\rho u_i$ for every $i \in \{0, \ldots, n-1\}$, and $s'\rho t'$, and
- if $t \overset{a}{\Longrightarrow}_\sharp t'$ with $(s, t') \notin \rho$, then there exists a path $s \overset{a_0}{\Longrightarrow} u_1 \overset{a_1}{\Longrightarrow} \ldots \overset{a_{n-1}}{\Longrightarrow} u_n \overset{a_n}{\Longrightarrow} s'$ with $n \geq 0$, $u_i\rho t$ for every $i \in \{0, \ldots, n-1\}$, and $s'\rho t'$.

Thus, a stutter bisimulation requires equivalent states to be equally labeled, and every outgoing transition in one system must be matched in the other system by a transition to an equivalent state, but allowing some transitions that are internal to the equivalence class of the source state. Note that action labels are not important here.

In our application, a stutter bisimulation $\rho \subseteq S \times S_\sharp$ between the original and the reduced system can inductively be defined as follows:

1. $s_0 \rho t_0$,
2. if $s\rho t$, $a \in A$, $s' \in S$, and $t' \in S_\sharp$ such that $s \overset{a}{\Longrightarrow} s'$, $t \overset{a}{\Longrightarrow}_\sharp t'$, and $\lambda(s') = \lambda_\sharp(t')$, then $s'\rho t'$, and
3. if $s\rho t$, $a \in A$, and $s' \in S$ such that $s \overset{a}{\Longrightarrow} s'$ and t has no $\overset{a}{\Longrightarrow}_\sharp$-successor, then $s'\rho t$.

This definition handles the following three cases: (1) it relates the initial states, (2) it relates states that are reachable from stutter-bisimilar states in both systems via the same machine instruction or via the same (non-blocked) IH, and (3) it relates a state that is reachable via some IH in the original system with the state in the reduced system where this IH is blocked.

The following arguments show that ρ is indeed a stutter bisimulation; details are omitted for lack of space. First, whenever $s\rho t$ with $s = (l, d) \in S$ and $t = (l_\sharp, d_\sharp) \in S_\sharp$, then $l = l_\sharp$ and $\lambda(l, d) = \lambda_\sharp(l_\sharp, d_\sharp)$. This is obvious in cases 1 and 2 of the definition of ρ, and also valid in 3 as the blocked IH returns to the same program location (implying $l = l_\sharp$), and must be invisible with respect to the atomic propositions (implying $\lambda(l, d) = \lambda_\sharp(l_\sharp, d_\sharp)$).

Second, the remaining requirements of a stutter bisimulation follow from the observation that, whenever $s\rho t$ (where $s \in S$ and $t \in S_\sharp$),

- if $s \overset{a}{\Longrightarrow} s'$ with $(s', t) \notin \rho$, then case 3 cannot apply as $s'\rho t$ otherwise. Hence, there exists $t' \in S_\sharp$ such that $t \overset{a}{\Longrightarrow}_\sharp t'$. For at least one of these states, it must be true that $\lambda(s') = \lambda_\sharp(t')$ (since $\lambda(s) \neq \lambda_\sharp(t)$ otherwise, contradicting $s\rho t$), and hence $s'\rho t'$;

– if $t \stackrel{a}{\Longrightarrow}_\sharp t'$ with $(s, t') \notin \rho$, then again case 2 must apply with $s \stackrel{a}{\Longrightarrow} s'$ and $s' \rho t'$.

The last step in our correctness proof consists of showing that both the original and the reduced transition system exhibit the same behavior with respect to non-terminating computations. Formally, a state $s \in S$ in a labeled transition system $(S, s_0, A, \Longrightarrow, P, \lambda)$ is called ρ-*divergent* with respect to an equivalence relation $\rho \subseteq S \times S$ if there exists an infinite path $s \stackrel{a_1}{\Longrightarrow} s_1 \stackrel{a_2}{\Longrightarrow} s_2 \stackrel{a_3}{\Longrightarrow} \ldots$ such that $s \rho s_i$ for all $i \geq 1$. The relation ρ is called *divergence-sensitive* if, for every $s_1 \rho s_2$, s_1 is ρ-divergent iff s_2 is ρ-divergent.

Again, it can be shown that the stutter bisimulation $\rho \subseteq S \times S_\sharp$ as defined above is also divergence-sensitive, the essential arguments being that non-terminating computations only occur in the form of cycles (as the state space is finite), and that our abstraction technique never completely blocks the execution of an IH in a loop, and therefore preserves divergence. This completes the proof that our abstraction technique is correct with respect to formulas in CTL*-X.

5   Case Study

This section describes a case study conducted with [MC]SQUARE using the IHER technique. We analyzed five programs for the ATMEL ATmega16 to evaluate the performance of our abstraction method. All programs were written by students during lab courses or diploma theses and have previously been used to evaluate the impact of other techniques developed for [MC]SQUARE. A more thorough description of the analyzed programs is given by Schlich [1]. Note that for all programs, delayed nondeterminism (DND) [2] is used, which affects state space sizes by delaying the instantiation of nondeterministic values until their concrete value is required. This way, [MC]SQUARE can handle programs of up to 4 billion (symbolic) states. The larger programs used in this case study could not be checked without DND.

The differences in state space sizes with and without IHER are presented in Table 1. It shows the numbers with and without dead variable reduction (DVR) enabled, which reduces state spaces by removing unused variables. Here, the formula **AG true** was checked as it requires the creation of the complete state space.

Two different versions of a controller for a powered window lift used in a car were analyzed, one of which containing defects caused by missing protection of shared variables, and a second one where those errors were fixed. Both programs consist of 290 lines of assembly code and use two interrupts and one timer. Depending on the applied static analysis techniques, the state space sizes are reduced by between 64% and 82%. The second program controls a fictive chemical plant. It consists of 225 lines of assembly code. One timer and two interrupts are used. The IHER technique reduced the state space by approx. 98%. The last program implements a four channel speed measurement with a CAN bus interface. It consists of 384 lines of assembly code. The state spaces were reduced by approx. 89%.

Table 1. Number of states stored by [MC]SQUARE

Without DVR	Default	Time [s]	IHER	Time [s]	Reduction
window_lift.elf (error)	316,334	6.25	64,164	7.32	80%
window_lift.elf (fixed)	129,030	2.81	23,852	6.04	82%
plant.elf (error)	123,699,464	3,428	2,161,624	43	98%
plant.elf (fixed)	75,059,765	1,956	1,327,715	25	98%
can.elf	147,259,483	3,917	16,187,483	392	89%
With DVR	**Default**	**Time [s]**	**IHER**	**Time [s]**	**Reduction**
window_lift.elf (error)	111,591	6.73	28,153	8.21	75%
window_lift.elf (fixed)	23,013	5.52	14,919	6.89	64%
plant.elf (error)	123,699,464	3,513	2,161,624	42	98%
plant.elf (fixed)	75,059,765	1,940	1,327,715	25	98%
can.elf	147,259,483	3,954	16,187,483	394	89%

These results show that the IHER abstraction technique greatly reduces state spaces for a number of different programs. This is still true in the presence of other abstraction techniques such as DVR, meaning that both can be combined. The magnitude of improvement depends on various factors such as the number of interrupts used, dependencies between IHs and instructions, dependencies between IHs, the overall structure of the program, and the property to be verified.

6   Related Work

In the past, much work has been carried out to limit the state explosion in model checking resulting from concurrent activities in a system. A prominent technique is partial order reduction (POR) [10,11,12], which tries to reduce the number of possible orderings of concurrent actions that need to be analyzed for model checking. This reduction is based on two important notions, namely, *independence* and *visibility*. Here, the first characterizes the commutativity of two actions, meaning that the execution of either of them does not disable the other and that executing both in any order always yields the same result. The second notion, visibility, refers to the property that the execution of an action does not affect the (in)validity of the formula to be checked. Together, both properties allow to reduce a transition system by only exploring a subset (the *ample set*) of all transitions enabled in a given state. A general overview of concepts related to POR is given by Valmari [13].

As pointed out in the introduction, our technique differs from POR in the following way. POR works in the context of concurrent threads while IHER works in the context of sequential programs and pseudo-parallelism introduced by IHs. Threads differ from IHs in that the interleaving between different threads is nondeterministic. For IHs only their occurrence is nondeterministic, that is, either they occur at a program location or they do not occur at a program location. Threads can be interrupted at any location since control can change nondeterministically between all threads. An IH, however, can interrupt the main program, but the main program cannot interrupt an IH. This means that an IH has to be executed completely until execution of the main program can continue. The same asymmetry applies in case that IHs can interrupt other IHs. Due to this asymmetry, we have to account for additional dependencies between the main program and IHs. In POR, all atomic actions are guaranteed to terminate. Since we represent IHs as atomic actions which can contain non-terminating loops, atomic actions are not guaranteed to terminate in our setting.

IHs could be modeled using threads. Techniques for converting interrupt-driven programs into equivalent programs using threads have been developed by Regehr and Cooprider [14]. This modeling can be done on source code level, but it involves some challenges. The peculiarities of interrupts vary between different microcontrollers. On some microcontrollers, IHs are non-interruptible while IHs can be interrupted on other microcontrollers. In some architectures interrupts have no priorities, in other architectures they have fixed or even dynamic priorities. This approach can, however, not be used for microcontroller assembly code as there is no thread model for microcontroller assembly code.

Kahlon et al. [15] developed an extension for partial order reductions using sound invariants. In their approach, the product graph of a concurrent system is iteratively refined, and statically unreachable nodes are removed. In contrast to our approach, only a context-insensitive static analysis is performed.

The notion of independent actions based on Lipton's theory of reduction [16] was introduced by Katz and Peled [17]. Our definition of dependencies between the main process and IHs can also be seen as an extension of Lipton's theory where, in addition to the dependencies that are induced by accesses to shared variables, also the control dependencies imposed by enabling and disabling interrupts are taken into account.

Recently, Elmas et al. [18] have described a proof calculus for static verification of concurrent programs using shared memory. In this approach, the concept of atomicity is used for computation of increased atomic code blocks, which are then, in contrast to our approach, verified sequentially.

A static analysis based on Petri nets to capture causal flows of facts in concurrent programs was proposed by Farzan and Madhusudan [19], but it only implements a restricted model of communication and synchronization compared to our setting. Another approach by Lal and Reps [20] adapts static analyses for sequential programs and extends them to work in a concurrent setting while our approach embodies specific analyses for concurrency. Other approaches, such as the work by Qadeer and Rehof [21] or Lal et al. [22], tackle the state explosion by imposing an upper bound on the number of context-switches, which is not possible in our setting.

7   Conclusion and Future Work

In this paper, we have presented a new abstraction technique called *interrupt handler execution reduction*, which is based on partial order reduction. It reduces state spaces by blocking the execution of interrupt handlers at certain program locations during model checking. It preserves the validity of CTL*-X and, as shown in the case study presented in Sect. 5, can significantly reduce state spaces. Two ingredients are needed for implementing this abstraction technique: a static analysis and a dynamic part executed during model checking. The static analysis determines program locations where interrupt handlers can be blocked. The model checking part then prevents the execution of the corresponding interrupt handlers at these program locations.

In the future, we want to improve the static analysis that is used for this abstraction technique. Currently, we rely on a coarse analysis of pointer variables. In many cases, our analysis has to over-approximate the set of possible address values. A more precise pointer analysis would improve the results of other static analyses such as live variable and reaching definitions analysis as well.

Another candidate for improvement is the refinement phase. From our point of view, there is no optimal static solution to this problem. We think that better heuristics can be found if termination of certain loops can be determined. Given this, we could postpone the execution of interrupt handlers beyond these loops.

References

1. Schlich, B.: Model Checking of Software for Microcontrollers. Dissertation, RWTH Aachen University, Aachen, Germany (June 2008)
2. Noll, T., Schlich, B.: Delayed nondeterminism in model checking embedded systems assembly code. In: Yorav, K. (ed.) HVC 2007. LNCS, vol. 4899, pp. 185–201. Springer, Heidelberg (2008)
3. Herberich, G., Noll, T., Schlich, B., Weise, C.: Proving correctness of an efficient abstraction for interrupt handling. In: Systems Software Verification (SSV 2008). ENTCS, vol. 217, pp. 133–150. Elsevier, Amsterdam (2008)
4. Emerson, E.A.: Handbook of Theoretical Computer Science. In: Handbook of Theoretical Computer Science, vol. B, pp. 995–1072. The MIT Press, Cambridge (1991)
5. Yorav, K., Grumberg, O.: Static analysis for state-space reductions preserving temporal logics. Formal Methods in System Design 25(1), 67–96 (2004)
6. Brauer, J., Schlich, B., Reinbacher, T., Kowalewski, S.: Stack bounds analysis for microcontroller assembly code. In: 4th Workshop on Embedded Systems Security (WESS 2009), Grenoble, France. ACM, New York (2009) (to appear)
7. Heljanko, K.: Model checking the branching time temporal logic CTL. Research Report A45, Helsinki University of Technology, Digital Systems Laboratory, Espoo, Finland (May 1997)
8. Browne, M., Clarke, E., Grumberg, O.: Characterizing finite kripke structures in propositional temporal logic. Theor. Comput. Sci. 59(1-2), 115–131 (1988)
9. van Glabbeek, R., Weijland, W.: Branching time and abstraction in bisimulation semantics. Journal of the ACM 43(3), 555–600 (1996)
10. Godefroid, P.: Using partial orders to improve automatic verification methods. In: Clarke, E., Kurshan, R.P. (eds.) CAV 1990. LNCS, vol. 531, pp. 176–185. Springer, Heidelberg (1991)
11. Holzmann, G.J., Peled, D.A.: An improvement in formal verification. In: Formal Description Techniques VII. IFIP International Federation for Information Processing, pp. 197–211. Springer, Heidelberg (1995)

12. Peled, D.: Ten years of partial order reduction. In: Y. Vardi, M. (ed.) CAV 1998. LNCS, vol. 1427, pp. 17–28. Springer, Heidelberg (1998)
13. Valmari, A.: The state explosion problem. In: Reisig, W., Rozenberg, G. (eds.) APN 1998. LNCS, vol. 1491, pp. 429–528. Springer, Heidelberg (1998)
14. Regehr, J., Cooprider, N.: Interrupt verification via thread verification. Electronic Notes in Theoretical Computer Science 174(9), 139–150 (2007)
15. Kahlon, V., Sankaranarayanan, S., Gupta, A.: Semantic reduction of thread interleavings in concurrent programs. In: Kowalewski, S., Philippou, A. (eds.) TACAS 2009. LNCS, vol. 5505, pp. 124–138. Springer, Heidelberg (2009)
16. Lipton, R.J.: Reduction: A method of proving properties of parallel programs. Communications of the ACM 18(12), 717–721 (1975)
17. Katz, S., Peled, D.: Defining conditional independence using collapses. Theoretical Computer Science 101(2), 337–359 (1992)
18. Elmas, T., Qadeer, S., Tasiran, S.: A calculus of atomic actions. In: Principles of Programming Languages (POPL 2009), Savanna, USA, pp. 2–15. ACM, New York (2009)
19. Farzan, A., Madhusudan, P.: Causal dataflow analysis for concurrent programs. In: Grumberg, O., Huth, M. (eds.) TACAS 2007. LNCS, vol. 4424, pp. 102–116. Springer, Heidelberg (2007)
20. Lal, A., Reps, T.: Reducing concurrent analysis under a context bound to sequential analysis. In: Gupta, A., Malik, S. (eds.) CAV 2008. LNCS, vol. 5123, pp. 37–51. Springer, Heidelberg (2008)
21. Qadeer, S., Rehof, J.: Context-bounded model checking of concurrent software. In: Halbwachs, N., Zuck, L.D. (eds.) TACAS 2005. LNCS, vol. 3440, pp. 93–107. Springer, Heidelberg (2005)
22. Lal, A., Touili, T., Kidd, N., Reps, T.: Interprocedural analysis of concurrent programs under a context bound. In: Ramakrishnan, C.R., Rehof, J. (eds.) TACAS 2008. LNCS, vol. 4963, pp. 282–298. Springer, Heidelberg (2008)

Diagnosability of Pushdown Systems

Christophe Morvan[1] and Sophie Pinchinat[2]

[1] Université Paris-Est,
INRIA Centre Rennes - Bretagne Atlantique
[2] IRISA, Campus de Beaulieu, 35042 Rennes, France

Abstract. Partial observation of discrete-event systems features a setting where events split into observable and unobservable ones. In this context, the diagnosis of a discrete-event system consists in detecting defects from the (partial) observation of its executions. Diagnosability is the property that any defect is eventually detected. Not surprisingly, it is a major issue in practical applications. We investigate diagnosability for classes of pushdown systems: it is undecidable in general, but we exhibit reasonably large classes of visibly pushdown systems where the problem is decidable. For these classes, we furthermore prove the decidability of a stronger property: the bounded latency, which guarantees the existence of a uniform bound on the respond delay after the defect has occurred. We also explore a generalization of the approach to higher-order pushdown systems.

1 Introduction

Absolute knowledge of the actual execution of a computer driven system is, in most settings, impossible. However, since typical systems emit information while interacting with their environment, deductions of their internal state can be performed on the basis of this *partial observation*.

From a mathematical point of view, a standard approach due to [17] uses a *discrete-event system* modeling (see, *e.g.*, [8]), provided with a partition of the event set into *observables* and *unobservables*. In this formal framework, *diagnosing* a system amounts to deducing, from its actual observation, the set I of its possible internal states, and to compare I with a distinguished subset of states P representing some property of the executions (for example the occurrence of a failure event). Diagnosing thereby brings about three different *verdicts*: the *negative* verdict when I does not meet P, the *positive* verdict when I lies in P, and the *inconclusive* verdict otherwise. The device which outputs the verdict is the *diagnoser*.

Building a diagnoser is not a difficult task, per se: it relies on classical power-set construction. For finite-state systems, it induces an unavoidable exponential blow-up [19], even for succinct representations [15]. Therefore on-the-fly computation of the diagnoser is a key techniques for effective methods. It incidentally offers an effective solution in infinite-state settings [18,3]. Whatever method is used for the diagnoser, a central question is whether the diagnoser will eventually

K. Namjoshi, A. Zeller, and A. Ziv (Eds.): HVC 2009, LNCS 6405, pp. 21–33, 2011.

detect any faulty execution (execution reaching P)? This property is the *diagnosability*, expressing intrinsic features of the system (together with P). Clearly, on-the-fly methods cannot apply, since diagnosability requires an exhaustive analysis of the model. PTIME decision procedures have been developed for finite-state systems [12,11]; non-diagnosability is NLOGSPACE-complete [15]. Also, SAT-solvers can be used for symbolic systems [9].

Beyond finite-state systems, very little exists in the literature on the diagnosis of discrete-event systems: [18] considered a timed systems setting, and established the equivalence between diagnosability and non-zenoness, yielding PSPACE-completeness. Petri nets have been studied in [20], where either classical techniques apply to finite nets (i.e. with a finite-state configuration graph), or approximation methods yield only semi-algorithms. Finally, [3] considered graph transformation systems, and developed a general procedure to compute the set of executions corresponding to a given observation. Notice that this approach does not provide any algorithm for the diagnosability whose statement universally quantifies over the set of observations. Surprisingly, diagnosis issues have never been addressed for pushdown systems, although acknowledged as good abstractions for the software model-checking of recursive programs [14]. Alternation-free (branching-time) μ-calculus, hence CTL, properties can be verified in EXPTIME [21], and fixed linear-time μ-calculus properties can be checked in PTIME [4]. In addition, partial observation of pushdown systems is simple to model since the class is closed under projection[1].

In this paper, we study diagnosability of pushdown systems (of arbitrary order) represented by (higher-order) pushdown automata. Diagnosability is shown undecidable in general, via a reduction of the emptiness problem for an intersection of context-free languages. In fact diagnosability requires concomitant properties that arbitrary classes of pushdown systems do not possess in general. Recently, Alur and Madhusudan introduced *visibly pushdown automata* [1] with adapted features to handle diagnosability. As we show here, arbitrary classes of visibly pushdown systems still do not yield decidability of the diagnosability, and our contribution precisely exhibits a sufficient condition. This condition correlates the observability of the system with its recursive structure: there must exists a pushdown description of the system, where accesses to the stack are observable. In this case, we adapt the non-diagnosability algorithm for finite-state systems developed in [11], yielding a PTIME upper bound; the NLOGSPACE lower bound for finite-state systems remains valid. The results on decidability are furthermore generalized to the higher order, by considering the higher-order visibly pushdown automata of [10]. We develop a k-EXPTIME algorithm for the class of k-order pushdown systems ($k \geq 2$).

As explained further in the paper, diagnosability guarantees, for each execution reaching P, a finite delay to detect it. However, it does not provide a uniform bound on these delays. The *bounded-latency* problem consists in deciding whether such a bound exists, and is fully relevant for practical applications. In the literature, bounded latency has been mainly investigated in the

---

[1] ε-closure of a pushdown automaton remains a pushdown automaton [2].

framework of finite-state systems, although it is a direct consequence of diagnosability. [16,22] refer to *the bound*, and [11] refer to *n-diagnosability*. Unexpectedly, to our knowledge, bounded latency has not been studied for infinite-state systems, for which diagnosability does not imply bounded latency.

In this paper, we consider the bounded-latency problem for pushdown systems. We show its decidability for families of first-order pushdown systems where diagnosability is already decidable (otherwise it does not make sense). For these families, bounded latency is equivalent to the finiteness of a language accepted by a pushdown autamaton. The latter problem is in PTIME [2]. Regarding higher-order pushdown systems, we conjecture undecidability of the bounded-latency problem. As for first-order pushdown systems, checking bounded latency amounts to checking finiteness of a higher-order pushdown language. For arbitrary higher-order pushdown language, the finiteness problem is still open, to our knowledge.

The paper is organized as follows. In Section 2 we define the diagnosability and the bounded-latency problems, and recall the classic results for finite-state systems. Pushdown systems are considered in Section 3, and handled in the core Section 4 of the contribution to study their diagnosability and bounded-latency problems. In Section 5, we consider higher-order pushdown systems.

2   Diagnosability and Bounded Latency

We first introduce some mathematical notations and definitions. Assume a fixed set E. We denote by 2^E its powerset, and by $\overline{B}$ the complement of a subset $B \subseteq E$. For any $k \in \mathbb{N}$, we write $[k] := \{1, 2, 3, \ldots, k\}$. Given an alphabet (a set of symbols) Σ, we write Σ^* and Σ^ω for the sets of finite and infinite words (sequences of symbols) over Σ respectively. We use the standard notation ε for the empty finite word, and we denote by $u, u', v, \ldots$ the typical elements of Σ^*, and by $w, w_1, \ldots$ the typical elements of Σ^ω. For $u \in \Sigma^*$, $|u|$ denotes the length of the word u.

Definition 2.1. A *discrete-event system* (DES) is a structure $\mathcal{S} = \langle \Sigma, S, s^0, \delta, Prop, [\![.]\!] \rangle$, where Σ is an alphabet, S is a set of states and $s^0 \in S$ is the initial state, $\delta : S \times \Sigma \to S$ is a (partial) transition function, and $Prop$ is a set of propositions and $[\![.]\!] : Prop \to 2^S$ is an interpretation of the propositions. An *execution* of $\mathcal{S}$ is a word $u = a_1 a_2 \ldots a_n \in \Sigma^*$ such that there exists a sequence of states $s_0, s_1, \ldots, s_n$ such that $s_0 = s^0$ and $\delta(s_{i-1}, a_i) = s_i$ for all $1 \leq i \leq n$. An execution u *reaches* a subset $S' \subseteq S$ whenever $\delta(s^0, u) \in S'$, by extending δ to $S \times \Sigma^*$. We naturally extend these definitions to *infinite* executions; in particular, an infinite execution $w \in \Sigma^\omega$ reaches S' if one of its prefixes reaches S'.

A proposition m *marks* the (elements of the) set $[\![m]\!]$, and an execution *reaches* m if it reaches $[\![m]\!]$.

We now give an overview on diagnosis. Diagnosis is about synthesis where one aims at constructing a device, a *diagnoser*, intended to work on-line together with the system. While the system executes, the diagnoser collects input data via sensors and outputs a verdict on the actual execution. In classic diagnosis, the sensors are not formally described, but instead simulated in a partial observation framework: the set of events Σ is partitioned into Σ_o and $\overline{\Sigma_o}$ composed of *observables* and *unobservables* respectively; words $\theta, \theta_1, \ldots$ over Σ_o are *observations*. The canonical projection of Σ onto Σ_o is written π_{Σ_o}, or π when Σ_o is understood; it extends to Σ^* by erasing unobservables in words. An execution u *matches* an observation θ whenever $\pi(u) = \theta$. Two executions u and u' are *indistinguishable* if they match the same observation.

Observations are the inputs of the diagnoser. Regarding the outputs, *faulty* executions of particular interest (as opposed to *safe* ones) are distinguished *a priori* by means of a proposition $f \in Prop$: an execution u is *faulty* if $\delta(s^0, u) \in \llbracket f \rrbracket$. Moreover, we require that $\llbracket f \rrbracket$ is a *trap*: $\delta(\llbracket f \rrbracket, a) \subseteq \llbracket f \rrbracket$, for every $a \in \Sigma$. This assumption means that we focus on whether some defect (a particular event or a particular pattern of events) has occurred in the past or not; we refer to [8] for a comprehensive exposition.

An instance of a diagnosis problem is a triplet composed of a DES, $\mathcal{S} = \langle \Sigma, S, s^0, \delta, Prop, \llbracket . \rrbracket \rangle$, an alphabet of observables, Σ_o, and a proposition, f. For technical reasons, we need to consider *information sets*: an information set I is the set of all states reached by a set of indistinguishable executions in $\Sigma^* \Sigma_o$. We write $\mathcal{I} \subseteq 2^S$ for the set of all information sets. Notice that $\{s_0\} \in \mathcal{I}$ and is associated to the empty observation. The associated diagnoser is a structure $\mathcal{D} := \langle \Sigma_o, \mathcal{I}, I^0, \hat{\delta}, diag \rangle$ whose states are either the initial state $I^0 := \{s^0\}$ or the transition function, $\hat{\delta} : \mathcal{I} \times \Sigma_o \to \mathcal{I}$, is the extension of δ to sets of states in a canonical way, and the output function $diag$ is defined as follows. Given a set $I \subseteq S$, three cases exist: (a) all states of I are marked by f; (b) no state is marked; and otherwise (c) where I is *equivocal*.

Formally,

$$diag : \mathcal{I} \to \{(a), (b), (c)\}$$
$$I \mapsto (a) \text{ if } I \subseteq \llbracket f \rrbracket, (b) \text{ if } I \cap \llbracket f \rrbracket = \emptyset, \text{ and } (c) \text{ otherwise}$$

By extension, an observation θ is *equivocal* if $\hat{\delta}(I^0, \theta)$ is equivocal, otherwise θ is *clear*; θ is *clearly-faulty* if it is clear and $\hat{\delta}(I^0, \theta)$ is in case (a). Since $I^0 = \{s^0\}$ is not equivocal, the empty observation is clear.

$\mathcal{D}$ may be infinite-state in general (if $\mathcal{S}$ is infinite-state). However, its computation can be avoided by simulating it on-the-fly, storing the current information set I, and updating this object on each observable step of the system. While the synthesis of the diagnoser is not necessary, analyzing its behaviour is crucial: in particular, because equivocalness (case (c)) precludes the instantaneous detection of a fault, latencies to react are tolerated.

Diagnosability is a qualitative property of the diagnoser which ensures a finite latency for any observation of a faulty execution; it corroborates the completeness of the diagnoser. From a quantitative point of view, the *bounded-latency* property ensure a uniform bound on the latencies. We develop these two notions.

In accordance with [17], we use the following definition (where the parameters Σ_o and f are understood).

Definition 2.2. A discrete-event system is *diagnosable* if every infinite observation of an infinite faulty execution has a clearly-faulty finite prefix.

Safe executions of diagnosable systems may have arbitrarily long equivocal observations as illustrated here with a system whose initial state p, and with the unobservable γ which leading to the marked state q. Since any faulty execution only yields an infinite observation with the clear prefix $a^n b$, the system is diagnosable, but the infinite observation a^ω of the the unique safe execution loops in the equivocal information set $\{p, q\}$.

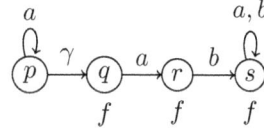

Lemma 2.3 [11]. A DES is not diagnosable w.r.t. the set of observables Σ_o and the proposition f if, and only if, there exist two indistinguishable infinite executions w_1 and w_2 such that w_1 reaches f while w_2 does not.

Notice that diagnosability considers only infinite executions that do not *diverge*, where an infinite executions diverges if it has an unobservable infinite suffix. In other words, we are only interested in fair behaviours of the system w.r.t. observability.

We now consider the *latency* of a diagnosable system as the minimal number of additional observation steps that is needed to detect a faulty execution.

Definition 2.4. Let $\mathcal{S} = \langle \Sigma, S, s^0, \delta, Prop, [\![.]\!] \rangle$ be a DES, Σ_o be an alphabet of observables, and $f \in Prop$ such that $[\![f]\!]$ is a trap. The *latency* of an execution u is defined by: $\ell(u) := \max \{|\vartheta|, \pi(u)\vartheta$ is not clearly-faulty$\}$ if u reaches f, and 0 otherwise.

$\mathcal{S}$ is *bounded-latency* if there exists $N \subset \mathbb{N}$ such that $\ell(u) \leq N$, for every execution u; the least such N is the *bounded-latency value*.

The bounded-latency value of the system above is 1: indeed, fix an observed execution u that reaches f and whose observation is not clearly-faulty (hence equivocal). This execution necessarily ends either in state q or in state r. If in q, the only sequence of observations ϑ such that $\pi(u)\vartheta$ is not clearly-faulty is $\vartheta = a$; therefore $\ell(u) = 1$. If in r, we have $\ell(u) = 0$.

Remark that a system is diagnosable if, and only if, $\ell(u)$ is a finite value, for every execution u, but not necessarily bounded. Therefore any bounded-latency system is diagnosable, but the converse does not hold in general.

The system depicted here is diagnosable when ι and γ unobservable and f (black) marking executions that contain the faulty event γ. Indeed, every maximal execution is finite, and its last event is ▶ if, and only if, γ has occurred. However, the system is not bounded-latency since arbitrarily many △'s can occur between γ and ▶.

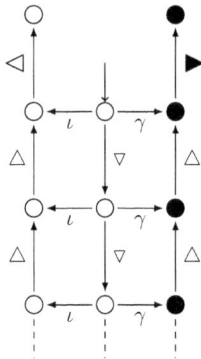

Since, diagnosability and bounded latency only depend on the set of executions of the system, one is allowed to decide these problems over a transformed system as long as executions are preserved.

For finite-state systems, it is easy to prove that diagnosability and bounded-latency properties coincide.

Theorem 2.5 [17,12,15]
For finite-state systems:
(i) Diagnosability is decidable in PTIME.
(ii) Non-diagnosability is NLOGSPACE-*complete.*

3   Pushdown Systems

We now investigate the case of pushdown systems where the picture is more involved. We recall that pushdown automata are finite-state machines that use a stack as an auxiliary data structure (see for example [2]); pushdown systems are derived as configuration graphs of pushdown automata and are infinite-state in general.

Definition 3.1. A *pushdown automaton (*PDA*)* is a structure $\mathcal{A} = (\Sigma, \Gamma, Q, q_0, F, \Delta)$ where Σ and Γ are finite alphabets of respectively *input* and *stack* symbols, Q is a finite set of states, $q_0 \in Q$ is the initial state, $F \subseteq Q$ is a set of final states, and $\Delta \subseteq Q \times (\Gamma \cup \{\varepsilon\}) \times (\Sigma \cup \{\varepsilon\}) \times Q \times \Gamma^*$ is the set of transitions.

We use $p, q, \ldots$ (resp. $X, Y, \ldots$, and $U, V, W, \ldots$) for typical elements of Q (resp. Γ, and Γ^*). Without loss of generality, we assume in *normal form*: (1) *pop* transitions of the form $(p, X, a, q, \varepsilon)$ pop the top symbol of the stack, (2) *push* transitions of the form $(p, \varepsilon, a, q, X)$ push a symbol on top of the stack, and (3) *internal* transitions of the form $(p, \varepsilon, a, q, \varepsilon)$ leave the stack unchanged. The PDA $\mathcal{A} = (\Sigma, \Gamma, Q, q_0, F, \Delta)$ is *deterministic* if: (1) $\forall (p, X, a) \in Q \times \Gamma \times \Sigma \cup \{\varepsilon\}$, there is at most one pair (q, V) such that $(p, X, a, q, V) \in \Delta$, and (2) $\forall (p, X) \in Q \times \Gamma$, if there exists (q, V) such that $(p, X, \varepsilon, q, V) \in \Delta$, then there is no triple $(q', a, V') \in Q \times \Sigma \times \Gamma^*$ such that $(p, X, a, q', V') \in \Delta$. An automaton $\mathcal{A}$ is *real-time* if $\Delta \subseteq Q \times (\Gamma \cup \{\varepsilon\}) \times \Sigma \times Q \times \Gamma^*$. A *configuration* of $\mathcal{A}$ is a word $qU \in Q\Gamma^*$; q is *the state* of the configuration. The *initial configuration* is $q_0\varepsilon$, and a configuration qU is *final* if $q \in F$. Transitions (between configurations) are elements of $Q\Gamma^* \times \Sigma \cup \{\varepsilon\} \times Q\Gamma^*$: there is a transition $(qU, a, q'U')$ whenever there exists $(q, X, a, q', V) \in \Delta$ with $U = WX$ and $U' = WV$. A finite *run* of $\mathcal{A}$ is a finite sequence $r = q_0U_0a_1q_1U_1a_2 \ldots a_nq_nU_n$ such that $U_0 = \varepsilon$ is initial, and

$(q_i U_i, a_i, q_{i+1} U_{i+1})$ is a transition, for all $0 \leq i < n$. We say that $a_1 a_2 \ldots a_n$ is *the word of* r, or that r is a *run on* $a_1 a_2 \ldots a_n$. The run is *accepting* if $q_n U_n$ is final.

The *language accepted* by $\mathcal{A}$ is $L(\mathcal{A}) \subseteq \Sigma^*$, the set of words $u \in \Sigma^*$ such that there is an accepting run on u.

Proposition 3.2 [2]. *Any* PDA *is equivalent to a real-time* PDA. *The construction is effective.*

PDA accept *context-free languages (*CF *languages)*, while *deterministic* PDA yield the proper subclass of *deterministic* CF languages, containing all regular languages. Moreover, CF languages are closed under union, concatenation, and iteration, but not under intersection, and their emptiness is decidable.

Proposition 3.3 [2]. *The emptiness problem of an intersection of deterministic* CF *languages is undecidable.*

We finally need to recall the following theorem.

Theorem 3.4 [4]. *The emptiness problem of a Büchi pushdown automaton is in* PTIME.

A *pushdown system (*PD *system)* $\mathcal{S}$ is the configuration graph of a real-time PDA $\mathcal{A} = (\Sigma, \Gamma, Q, q_0, F, \Delta)$, which *represented* $\mathcal{S}$. Notice that the set F of $\mathcal{A}$ is irrelevant for $\mathcal{S}$. However, using standard techniques, the statement that $[\![f]\!]$ is a regular set of configurations in $\mathcal{S}$ can be transformed into $[\![f]\!] = F\Gamma^*$ (see appendix for details).

By Proposition 3.3, PD systems are not closed under product (usual synchronous product), which causes limitations in effective methods for their analysis, and in particular regarding diagnosis (Sect. 4). We therefore consider more friendly sub-classes of PDA: the *visibly pushdown automata* [1].

Visibly pushdown automata are PDA with restricted transition rules: whether a transition is push, pop, or internal depends only on its input letter.

Definition 3.5. A *visibly pushdown automaton (*VPA*)* is a pushdown automaton $\mathcal{A} = (\Sigma, \Gamma, Q, q_0, F, \Delta)$, where $\bot \in \Gamma$ is a special bottom-stack symbol, and whose input alphabet and transition relation are partitioned into $\Sigma := \Sigma_{push} \cup \Sigma_{pop} \cup \Sigma_{int}$, where Σ_{int} is the *internal alphabet*, and $\Delta := \Delta_{push} \cup \Delta_{pop} \cup \Delta_{int}$ respectively, with the constraints that $\Delta_{push} \subseteq Q \times \{\varepsilon\} \times \Sigma_{push} \times Q \times (\Gamma \setminus \{\bot\})$, $\Delta_{pop} \subseteq Q \times \Gamma \times \Sigma_{pop} \times Q \times \{\varepsilon\}$, and $\Delta_{int} \subseteq Q \times \{\varepsilon\} \times \Sigma_{int} \times Q \times \{\varepsilon\}$.

A $[\Sigma_{int}]$-VP *language* is a language accepted by some VPA whose internal alphabet is Σ_{int}.

Theorem 3.6 [1]. *(a) Any* VPA *is equivalent to a deterministic* VPA *over the same alphabet. The construction is effective.*
(b) Any family of VP *languages with a fixed partition* $\Sigma_{push}, \Sigma_{pop}, \Sigma_{int}$ *of the input alphabet is a Boolean algebra. In particular the synchronous product* $\mathcal{A}_1 \times \mathcal{A}_2$ *of* VPA *is well-defined.*

We now turn to the projection operation on languages with respect to a sub-alphabet as a central operation for partial observation issues; we recall that the class of CF languages is projection-closed, whereas VP languages are not; more precisely,

Proposition 3.7
(i) Any CF language is the projection of a $[\emptyset]$-VP language, and this is effective.
(ii) The projection of a $[\Sigma_{int}]$-VP language onto Σ'^*, with $\overline{\Sigma'} \subseteq \Sigma_{int}$, is a $[\Sigma_{int}]$-VP language (with $\Sigma_{push}, \Sigma_{pop}$ fixed). The construction is effective.

4   Diagnosability and Bounded Latency of PD Systems

We show that diagnosability of arbitrary deterministic PD systems is undecidable. Next, we focus on VP systems whose diagnosability is also undecidable in general, unless unobservable transitions leave the stack unchanged.

Theorem 4.1. *Diagnosability of deterministic PD systems is undecidable.*

This theorem is a corollary of Proposition 3.3 and the following construction together with Lemma 4.2. Let $\mathcal{A}_1$ and $\mathcal{A}_2$ be two deterministic PDA over Σ_1 and Σ_2 respectively, and let $\Sigma = \Sigma_1 \cup \Sigma_2 \cup \{\iota_1, \iota_2, \#\}$, with fresh symbols $\#$, ι_1 and ι_2.

For $i = 1, 2$, let $\mathcal{A}_i^\#$ be a deterministic PDA which accepts $L(\mathcal{A}_i)\#\Sigma^*$, the set of words $u\#v$ where $u \in L(\mathcal{A}_i)$. Let $\mathcal{A}_1^\# \oplus \mathcal{A}_2^\#$ be the PDA depicted on the right. Mark all configurations of $\mathcal{A}_1^\# \oplus \mathcal{A}_2^\#$ whose state is in $\mathcal{A}_1^\#$ by f; $[\![f]\!]$ is a regular set and a trap, by construction. Notice that $\mathcal{A}_1^\# \oplus \mathcal{A}_2^\#$ is deterministic.

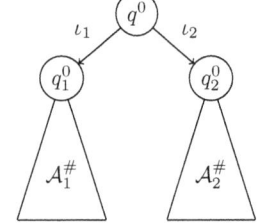

Lemma 4.2. The PD system $\mathcal{S}$ represented by $\mathcal{A}_1^\# \oplus \mathcal{A}_2^\#$ is diagnosable w.r.t. $\Sigma \setminus \{\iota_1, \iota_2\}$ and f if, and only if, $L(\mathcal{A}_1) \cap L(\mathcal{A}_2) = \emptyset$.

Indeed, consider $w_1 := \iota_1 u\#^\omega$ indistinguishable from $w_2 := \iota_2 u\#^\omega$ with $u \in L(\mathcal{A}_1) \cap L(\mathcal{A}_2)$. Thus w_1 reaches f but w_2 does not. Apply Lemma 2.3 to conclude. Reciprocally, if $\mathcal{S}$ is not diagnosable, then by Lemma 2.3, there exist indistinguishable infinite executions w_1 and w_2 such that only w_1 reaches f; necessarily, $w_1 = \iota_1 u\#w$ and $w_2 = \iota_2 u\#w$ for some u, entailing $u \in L(\mathcal{A}_1) \cap L(\mathcal{A}_2)$, which concludes the proof.   □

Theorem 4.3. *(a) Diagnosability of VP systems is undecidable.*
(b) Diagnosability w.r.t. a set of observables Σ_o and a proposition f is decidable in PTIME over any class of $[\Sigma_{int}]$-VP systems whenever $\overline{\Sigma_o} \subseteq \Sigma_{int}$ and f marks a regular set of configurations.

Proof. Point (a) is an immediate corollary of the undecidability of diagnosability for PD systems (Theorem 4.1) and the fact that any CF language is the projection of some VP language (Proposition 3.7 Point (i)). For Point (b) of Theorem 4.3, let $\mathcal{S}$ be a VP system represented by a deterministic $[\Sigma_{int}]$-VPA $\mathcal{A} = (\Sigma, \Gamma, Q, q_0, F, \Delta)$, and consider an alphabet of observables Σ_o such that $\overline{\Sigma_o} \subseteq \Sigma_{int}$, and a proposition f which marks a regular set of configurations of $\mathcal{A}$. We sketch an algorithm to decide the diagnosability of $\mathcal{S}$ w.r.t. Σ_o and f. The proposed method extends the solution of [11] for finite-state systems.

Consider the (non-deterministic) $[\Sigma_{int} \setminus \overline{\Sigma_o}]$-VPA $\pi(\mathcal{A}) \times \pi(\mathcal{A})$ (over the stack alphabet $\Gamma \times \Gamma$), obtained by Σ_o-projecting $\mathcal{A}$ and by building the standard product of VPA [1].

Lemma 4.4. The VPA $\pi(\mathcal{A}) \times \pi(\mathcal{A})$ with initial state (q_0, q_0) and final states $F \times \overline{F}$ accepts the equivocal observations.

Note that for $\pi(\mathcal{A}) \times \pi(\mathcal{A})$, an infinite run remaining in the set of configurations $(F \times \overline{F})(\Gamma \times \Gamma)^*$ denotes an infinite observation which has no clear prefix. By Lemma 2.3, this equivalently rephrases as "the system is not diagnosable". Now, the existence of such a run is equivalent to check the non emptiness of the Büchi automaton whose structure is $\pi(\mathcal{A}) \times \pi(\mathcal{A})$ and whose accepting states are all elements of $(F \times \overline{F})$ (use the fact that $[\![f]\!]$ is a trap). By Theorem 3.4, this can be decided in NLOGSPACE.   □

We now establish that for the classes of PD systems that yield effective methods to answer diagnosability problems, bounded latency is also decidable.

Theorem 4.5. *Given a* $[\Sigma_{int}]$-*VP system* $\mathcal{S}$, *an observation alphabet* Σ_o *with* $\overline{\Sigma_o} \subseteq \Sigma_{int}$, *and a proposition* f *which marks a regular set of configurations, it is decidable in* PTIME *whether* $\mathcal{S}$ *is bounded latency or not. Furthermore, the bound can be effectively computed.*

Proof. Without loss of generality, we can assume $\mathcal{S}$ diagnosable (which is decidable by the hypothesis and Theorem 4.3), otherwise it is not bounded-latency.

Let the deterministic $[\Sigma_{int}]$-VPA $\mathcal{A}$ represent $\mathcal{S}$. Derive from the VPA $\pi(\mathcal{A}) \times \pi(\mathcal{A})$ the (non-deterministic) PDA $\mathcal{A}'$ as follows: re-label with ε all transitions leaving states in $\overline{F} \times \overline{F}$, remove all transitions leaving states in $F \times F$, let (q_0, q_0) be the initial state, and let $F \times F$ be the final states. As such, $\mathcal{A}'$ accepts the words ϑa $(a \in \Sigma_o)$ where for some execution u that reaches f, $\pi(u)\vartheta a$ is clearly-faulty but $\pi(u)\vartheta$ is not. By Definition 2.4, $L(\mathcal{A}')$ is finite (which is decidable in PTIME [2]) if, and only if, $\mathcal{S}$ is bounded-latency; if finite, the value is $\max\{|\vartheta| \mid \vartheta \in L(\mathcal{A}')\Sigma^{-1}\}$.[2]   □

5   Extension to Higher-Order Pushdown Systems

Higher-order pushdown automata [13] extend PDA and reach context-sensitive languages. We only sketch their definition, following [7].

---

[2] We use the standard notation $U\Sigma^{-1}$ to denote the set of words v such that $v.a \in U$ for some $a \in \Sigma$.

Let Γ be a stack alphabet. For any integer $k \geq 1$, k *level* stacks, or shortly *k-stacks*, (over Γ) are defined by induction: A 1-stack is of the form $[U]_1$, where $U \in \Gamma^*$, and the empty stack is written $[]_1$; 1-stacks coincide with stacks of PDA. For $k > 1$, a k-stack is a finite sequence of $(k-1)$-stacks; the empty k-stack is written $[]_k$. An *operation of level k* acts on the topmost k-stack of a $(k+1)$-stack; operations over stacks (of any level) preserve their level. Operations of level 1 are the classical $push_X$ and pop_X, for all $X \in \Gamma$: $push_X([U]_1) = [UX]_1$ and $pop_X([UX]_1) = [U]_1$. Operations of level $k > 1$ are $copy_k$ and $\overline{copy}_k$, and act on $(k+1)$-stacks as follows ($S_1, \ldots, S_n$ are k-stacks).

$$copy_k([S_1, \ldots, S_n]_{k+1}) := [S_1, \ldots, S_n, S_n]_{k+1}$$
$$\overline{copy}_k([S_1, \ldots, S_n, S_n]_{k+1}) := [S_1, S_2, \ldots, S_n]_{k+1}$$

Any operation ρ of level k extends to arbitrary higher level stacks according to: $\rho([S_1, \ldots, S_n]_\ell) = [S_1, \ldots, \rho(S_n)]_\ell$, for $\ell > k+1$.

A *higher-order pushdown automaton (HPDA) of order k* is a structure $\mathcal{A} = (\Sigma, \Gamma, Q, q_0, F, \Delta)$ like a PDA, but where Δ specifies transitions which affect operations on the k-stack of the automaton. We refer to [7] for a comprehensive contribution on the analysis of HPDA; following this contribution, a set of configurations is *regular* whenever the sequences of operations that are used to reach the set form a regular language, in the usual sense. *Higher-order pushdown systems (HPDS)* are configuration graphs of HPDA. By Theorem 4.1, their diagnosability is undecidable. However, similarly to first-order PD systems, *higher-order VPA (HVPA)* can be considered [10].

A *k-order VPA* has $(2k+1)$ sub-alphabets Σ_{push}, Σ_{pop}, Σ_{int}, Σ_{copy_r}, and $\Sigma_{\overline{copy}_r}$, where $r \in [k]$, each of which determines the nature (*e.g.* push, pop, internal, $copy_r$, $\overline{copy}_r$) of the transitions on its symbols. Transitions on elements of Σ_{int} leave the stacks of any level unchanged. According to [10], HVPA are neither closed under concatenation, nor under iteration, and cannot be determinized; they are however closed under intersection.

Proposition 5.1. *The projection onto Σ'^* of a k-order VP language with internal alphabet Σ_{int} is a k-order VP language, provided $\overline{\Sigma'} \subseteq \Sigma_{int}$.*

Proof. The proof of Proposition 3.7 easily adapts here. Let L be a k-order VP language accepted by the k-order HVPA $\mathcal{A} = (\Sigma, \Gamma, Q, q_0, F, \Delta)$. We again write $p \Rightarrow p'$ whenever there exists $(p, \varepsilon, a, p', \varepsilon) \in \Delta_{int}$ with $a \in \overline{\Sigma_o}$.

The HVPA $\pi(\mathcal{A})$ which accepts $\pi(L)$ is obtained by adding new transitions, and by letting $p \in F'$ if $p \Rightarrow^* p'$, for some $p' \in F$. The transitions in Δ' are obtained by replacing, in a transition of Δ, the origin state p by the state r, provided $r \Rightarrow p$ in $\mathcal{A}$. Notice that $\Delta \subseteq \Delta'$.

This construction is correct in the sense that $L(\pi(\mathcal{A})) = \pi(L)$. $\square$

Theorem 5.2. *For any class of k-order VP systems with the sub-alphabets Σ_{push}, Σ_{pop}, Σ_{int}, Σ_{copy_r}, and $\Sigma_{\overline{copy}_r}$ $(r \in [k])$, diagnosability w.r.t. the set of observables Σ_o and the proposition f is decidable in k-EXPTIME, whenever $\overline{\Sigma_o} \subseteq \Sigma_{int}$ (the internal alphabet) and f marks a regular set of configurations.*

Proof. Let $\mathcal{S}$ be a k-order VP system represented by $\mathcal{A} = (\Sigma, \Gamma, Q, q_0, F, \Delta)$. By Proposition 5.1, $\pi(\mathcal{A})$ is a k-order VPA, and Lemma 4.4 for first-order VP system can be easily adapted.

Lemma 5.3. *The non-deterministic k-order VPA $\pi(\mathcal{A}) \times \pi(\mathcal{A})$ with initial state (q_0, q_0) and final states $F \times \overline{F}$ accepts the equivocal observations.*

Assuming the VPA $\mathcal{A}$ is a k-order pushdown automaton, so is the VPA $\pi(\mathcal{A}) \times \pi(\mathcal{A})$. As in the proof of Theorem 4.3, checking diagnosability amounts to decide the non emptiness of the language accepted by the Büchi k-order pushdown automaton $\pi(\mathcal{A}) \times \pi(\mathcal{A})$ with accepting states in $F \times \overline{F}$. According to [5], this is decidable in k-EXPTIME, but the lower bound is still an open question. $\square$

Regarding the bounded-latency problem, Theorem 4.5 does not easily extend to HVP systems. Indeed, in the proof of this theorem, deciding the finiteness of a CF language (namely $L(\mathcal{A}')$ page 29) is a key point, and fortunately this is decidable: the standard decision procedure makes the assumption that the automaton to represent the language is real-time, which is always possible for CF languages using an effective method. If we were able to restrict to real-time HPDA, we would have a similar result since one can show the following.

Theorem 5.4. *The finiteness of a real-time HPD language is decidable.*

Nevertheless, it is an open question whether arbitrary higher-order pushdown languages are real-time or not; in fact, [6] conjectures they are not. At the moment, deciding the finiteness of an arbitrary HPD language is a difficult question, and so is the bounded-latency property of a higher-order pushdown system, as resolving the latter problem solves the former.

Proposition 5.5. *Let $\mathcal{L}$ be a class of higher-order pushdown languages which is closed under concatenation and union. For each $L \in \mathcal{L}$, there exists a DES $\mathcal{S}_L$ such that $\mathcal{S}_L$ is bounded latency if, and only if, L is finite.*

Proof. Assume $L \in \mathcal{L}$ with alphabet Σ. The set of events of $\mathcal{S}_L$ is $\Sigma \cup \{\iota_1, \iota_2, \#, \$\}$, with fresh symbols ι_1, ι_2, $\#$, and $\$$. $\mathcal{S}_L$ has two components $L\#$ and $L\$$ (see figure next page). By construction, the set of executions of $\mathcal{S}_L$ is in $\mathcal{L}$. By letting ι_1 and ι_2 be unobservable, and f mark the configurations of the $L\$$ component, $\mathcal{S}_L$ is diagnosable. 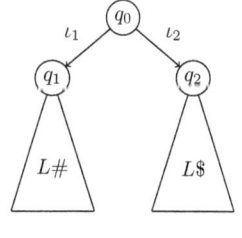 Indeed, event $\#$ or event $\$$ always eventually occur along any execution, revealing the actual running component of the system. It is easy to verify that $\mathcal{S}_L$ is bounded-latency if, and only if, L finite. $\square$

As a consequence, Proposition 5.5 considerably lessens hopes to decide the bounded latency problem for arbitrary HVP systems. We nevertheless exhibit cases where the problem can sometimes be answered.

Consider a HVP system represented by the HVPA $\mathcal{A} = (\Sigma, \Gamma, Q, q_0, F, \Delta)$. By Lemma 5.3, the real-time HPDA $\pi(\mathcal{A}) \times \pi(\mathcal{A})$ (with initial state (q_0, q_0) and final states $F \times \overline{F}$) accepts the set Υ of equivocal observations, whose finiteness is decidable by Theorem 5.4. We consider the possible cases:

If Υ is finite, then the system is bounded-latency and the bound is $\max\{|\vartheta| \mid \exists \theta \in \Upsilon, \theta\vartheta$ is not clearly-faulty$\}$.

Otherwise, we inspect the set $\mathcal{C}$ of configurations reached by Υ, which by [7] is a regular set (that can be effectively computed). Now, decide whether $\mathcal{C}$ is finite or not.

If $\mathcal{C}$ is finite, for each configuration $C \in \mathcal{C}$, build the real-time HPDA $\mathcal{A}_C$ as follows: (1) cut in the automaton $\pi(\mathcal{A}) \times \pi(\mathcal{A})$ every transitions that leaves $F \times F$, (2) set C as the initial configuration, and (3) $F \times F$ as the final states. Since $L(\mathcal{A}_C)$ is a real-time HPDA its finiteness is decidable. If every $L(\mathcal{A}_C)$ is finite (which can be check since $\mathcal{C}$ is finite), then the system is bounded-latency and the bound is $\max\{|\vartheta| \mid \vartheta \in (\cup_{C \in \mathcal{C}} L(\mathcal{A}_C)).\Sigma^{-1}\}$.

If $\mathcal{C}$ is infinite, nothing can be inferred.

References

1. Alur, R., Madhusudan, P.: Visibly pushdown languages. In: STOC 2004, pp. 202–211. ACM, New York (2004)
2. Autebert, J.-M., Berstel, J., Boasson, L.: Context-free languages and pushdown automata. In: Handbook of formal languages, vol. 1, pp. 111–174. Springer, Heidelberg (1997)
3. Baldan, P., Chatain, T., Haar, S., König, B.: Unfolding-based diagnosis of systems with an evolving topology. In: van Breugel, F., Chechik, M. (eds.) CONCUR 2008. LNCS, vol. 5201, pp. 203–217. Springer, Heidelberg (2008)
4. Bouajjani, Esparza, Maler: Reachability analysis of pushdown automata: Application to model-checking. In: Mazurkiewicz, A., Winkowski, J. (eds.) CONCUR 1997. LNCS, vol. 1243, pp. 135–150. Springer, Heidelberg (1997)
5. Cachat, T., Walukiewicz, I.: The complexity of games on higher order pushdown automata. CoRR, abs/0705.0262 (2007)
6. Carayol, A.: Notions of determinism for rational graphs. Private Communication (2001)
7. Carayol, A.: Regular sets of higher-order pushdown stacks. In: Jedrzejowicz, J., Szepietowski, A. (eds.) MFCS 2005. LNCS, vol. 3618, pp. 168–179. Springer, Heidelberg (2005)
8. Cassandras, C.G., Lafortune, S.: Introduction to Discrete Event Systems. Kluwer Academic Publishers, Dordrecht (1999)
9. Grastien, A., Anbulagan, Rintanen, J., Kelareva, E.: Diagnosis of discrete-event systems using satisfiability algorithms. In: AAAI, pp. 305–310. AAAI Press, Menlo Park (2007)
10. Illias, S.: Higher order visibly pushdown languages, Master's thesis, Indian Institute of Technology, Kanpur (2005)
11. Jéron, T., Marchand, H., Pinchinat, S., Cordier, M.-O.: Supervision patterns in discrete event systems diagnosis. In: 8th Workshop on Discrete Event Systems, Ann Arbor, Michigan, USA (July 2006)
12. Jiang, S., Huang, Z., Chandra, V., Kumar, R.: A polynomial time algorithm for diagnosability of discrete event systems. IEEE Transactions on Automatic Control 46, 1318–1321 (2001)

13. Maslov, A.: Multilevel stack automata. Problems of Information Transmission 12, 38–43 (1976)
14. Ong, C.-H.L.: Hierarchies of infinite structures generated by pushdown automata and recursion schemes. In: Kučera, L., Kučera, A. (eds.) MFCS 2007. LNCS, vol. 4708, pp. 15–21. Springer, Heidelberg (2007)
15. Rintanen, J.: Diagnosers and diagnosability of succinct transition systems. In: Veloso, M.M. (ed.) IJCAI, pp. 538–544 (2007)
16. Sampath, M., Sengupta, R., Lafortune, S., Sinaamohideen, K., Teneketzis, D.: Diagnosability of discrete event systems. IEEE Transactions on Automatic Control 40, 1555–1575 (1995)
17. Sampath, M., Sengupta, R., Lafortune, S., Sinaamohideen, K., Teneketzis, D.: Failure diagnosis using discrete event models. IEEE Transactions on Control Systems Technology 4, 105–124 (1996)
18. Tripakis, S.: Fault diagnosis for timed automata. In: Damm, W., Olderog, E.-R. (eds.) FTRTFT 2002. LNCS, vol. 2469, pp. 205–224. Springer, Heidelberg (2002)
19. Tsitsiklis, J.N.: On the control of discrete event dynamical systems. Mathematics of Control Signals and Systems 2, 95–107 (1989)
20. Ushio, T., Onishi, I., Okuda, K.: Fault detection based on Petri net models with faulty behaviors. In: IEEE International Conference on Systems, Man, and Cybernetics, vol. 1, pp. 113–118 (1998)
21. Walukiewicz, I.: Model checking ctl properties of pushdown systems. In: Kapoor, S., Prasad, S. (eds.) FST TCS 2000. LNCS, vol. 1974, pp. 127–138. Springer, Heidelberg (2000)
22. Yoo, T.-S., Lafortune, S.: Polynomial-time verification of diagnosability of partially-observed discreteevent systems. IEEE Transactions on Automatic Control 47, 1491–1495 (2002)

Functional Test Generation with Distribution Constraints

Anna Moss and Boris Gutkovich

Intel Corporation, Haifa, Israel
{anna.moss,boris.gutkovich}@intel.com

Abstract. In this paper, we extend the Constraint Programming (CP) based functional test generation framework with a novel concept of distribution constraints. The proposed extension is motivated by requirements arising in the functional validation field, when a validation engineer needs to stress an interesting architectural event following some special knowledge of design under test or a specific validation plan. In such cases there arises the need to generate a sequence of test instructions or a collection of tests according to user-given distribution requirements which specify desired occurrence frequencies for interesting events. The proposed extension raises the expressive power of the CP based framework and allows specifying distribution requirements on a collection of Constraint Satisfaction Problem (CSP) solutions. We formalize the notion of distribution requirements by defining the concept of distribution constraints. We present two versions of problem definition for CP with distribution constraints, both of which arise in the context of functional test generation. The paper presents algorithms to solve each of these two problems. One family of the proposed algorithms is based on CP, while the other one makes use of both CP and the linear programming (LP) technology. All of the proposed algorithms can be efficiently parallelized taking advantage of the multi core technology. Finally, we present experimental results to demonstrate the effectiveness of proposed algorithms with respect to performance and distribution accuracy.

1 Introduction

A major step in processor design cycle is design verification on the register transfer level (RTL). One of the commonly used approaches to this task is simulation-based validation. In this approach, the design under test (DUT) is examined against a large amount of functional tests, which exercise numerous execution scenarios in order to expose potential bugs. Due to size and complexity of modern designs it is not feasible to exercise the DUT on all possible test scenarios. Instead, a common approach is to develop a representative sample of possible tests. The latter is obtained by generating so called directed random tests, which are driven by constraints to express test intention yet randomization is applied to unconstrained test components. In this approach, a single constraint specification is associated with a huge collection of tests satisfying the specified constraints (solution space) and randomization is applied to sample this solution space. The work of validation engineers on developing test suites is facilitated by the use of automated test generation tools. A powerful means in performing

K. Namjoshi, A. Zeller, and A. Ziv (Eds.): HVC 2009, LNCS 6405, pp. 34–51, 2011.
© Springer-Verlag Berlin Heidelberg 2011

automated functional test generation is the CP technology. In particular, constraint modeling provides the capability to declaratively describe the DUT specification which defines a valid test as well as to describe a specific test scenario. Moreover, advanced CP algorithms can be used by automated test generation tools to produce tests that answer the architecture and scenario requirements. The examples of CP technology applications to the functional test generation task can be found in [1], [2], [3], [4].

The quality of sampling of the solution space is one of the major factors in judging the quality of a test generation tool. A failure to provide a collection of tests giving a good representation of the entire solution space and its required subspaces would translate to coverage holes in design verification with the possible implication of undiscovered bugs. A common requirement related to sampling of the solution space is that generated tests should be distributed uniformly at random over the space of all tests satisfying the given constraints. Providing such uniform distribution is a difficult problem in constraint solving. Some research has been done on this problem in the context of SAT [5] as well as in the context of CP [6]. However, the complexity of the algorithm proposed in the latter work makes it hard to use in CSP problems arising in the functional test generation task. Another approach to address the quality of solution sampling is to define diversity measurements, e.g. average Hamming distance, on a collection of solutions and generate a collection of solutions maximizing the diversity between solutions in the collection [7].

From the discussion above it follows that CP based functional test generation can be seen in particular as the task of generating a collection of multiple CSP solutions. Functional validation domain provides use cases where requirements need to be applied to the collection of generated solutions on the whole rather than to individual solutions. The uniformity of solution space sampling and the diversity of solutions in the generated collection are examples of such implicit requirements. However, there are use cases where *explicit* user given requirements need to be applied to a collection of solutions. For example, a validation engineer might wish some interesting architectural event to occur in some given fraction of test instructions without binding the event to a specific subset of instructions in the test. On the other hand, the traditional CP based framework allows expressing constraints only on individual CSP solutions but not on a collection of solutions. In this paper we propose to extend the CP paradigm with the notion of *distribution constraints* that apply to a collection of solutions. Such extension raises the expressive power of the CP based framework to answer the mentioned above needs of the functional validation. From the aspect of modeling, the extended framework should provide a user the capability to formulate distribution requirements for desired properties in a collection of multiple solutions. From the aspect of search, it should provide efficient algorithms for solution space sampling resulting in a collection of solutions satisfying given distribution requirements.

To the best of our knowledge, very little research has been done in relation to distribution constraints. Larkin [8] proposed a very general definition of distribution requirements, allowing them to be specified as an arbitrary Bayesian network, and presented algorithms for sampling satisfying such requirements. However, these algorithms are exponential in the number of CSP variables, which can reach thousands in a typical CSP corresponding to functional test generation. This makes the algorithms inapplicable to the functional test generation problem. Another work in this direction has been done by one of the authors of this paper [9]. There, a more restrictive definition of distribution requirements was introduced. We observe that distribution constraints paradigm is closely related to another concept in test generation known as *biasing*. Biasing is a (user-given) heuristic for selecting values of variables. For example, each possible value could be selected at random with a given probability. In certain cases, biasing and its extensions can serve as an approximation to solution space sampling subject to given distribution requirements. This approach was taken in [9] to provide algorithms for distribution constraint satisfaction. While the performance of these algorithms is good, they provide no guarantees with respect to the distribution accuracy, and while giving a reasonable approximation in some cases, may have poor accuracy in the others. In addition, the results of these algorithms are strongly influenced by algorithm implementation specifics, like variable ordering.

In this paper, we present a refined formal definition of distribution constraints. The proposed definition is less general than that of an arbitrary Bayesian network, however provides sufficient expressiveness for specifying distribution requirements in the functional test generation task. We also define a generalization of the distribution constraint concept which we refer to as *conditional* distribution constraints, to allow the capability of expressing another common type of distribution requirements. The paper presents two versions of the problem of solution space sampling subject to distribution constraints, both of which arise in the context of functional test generation. For each of these two problems, we propose algorithms for finding a sampling satisfying given distribution constraints. These algorithms are both efficient in terms of performance, thus improving on results in [8], and on the other hand provide high distribution accuracy and eliminate dependency on search algorithm specifics, thus overcoming the drawbacks of the results in [9]. The first type of the proposed algorithms is based on the CP search combined with a parallelization scheme, while the other type makes use of the LP technology. We provide experimental results to demonstrate efficiency and distribution accuracy of the proposed algorithms.

The rest of the paper is organized as follows. In section 2 we provide the background required for presentation of our results. Section 3 presents definitions of distribution constraints as well as conditional distribution constraints. In Section 4 we define two problems of sampling solution space subject to distribution constraints. Section 5 describes algorithms for each of the presented problem definitions. Section 6 demonstrates experimental results. We conclude in Section 7 with the summary of the presented results.

2 Background

For the sake of completeness, in this section we provide the CP and LP background required to facilitate the presentation of the rest of this paper. An in-depth survey of the traditional CP can be found in [10] whereas extended CP frameworks are surveyed in [11]. LP related definitions and theory can be found in [12].

The CP paradigm comprises the modeling of a problem as a CSP, constraint propagation, search algorithms, and heuristics. A CSP is defined by:

- a set of constrained variables. Each variable is associated with a (finite) domain defined as a collection of values that the variable is allowed to take;
- a set of constraints. A constraint is a relation defined on a subset of variables which restricts the combinations of values that the variables can take simultaneously.

A *solution* to a CSP is an assignment of values to variables so that each variable is assigned a value from its domain and all the constraints are satisfied.

The constraints referred to in the classical CSP definition above are also known as *hard* constraints, in the sense that any feasible solution of the CSP must satisfy these constraints. A number of extended CP frameworks have been proposed that relax the notion of a constraint to better suit some real world problems. One of the proposed relaxations of a classical hard constraint is a *soft* constraint which is allowed to be violated if it cannot be satisfied. CP with soft constraints is known as Partial Constraint Satisfaction [13]. Different criteria have been proposed to measure the quality of solutions for a CSP with soft constraints and many algorithms have been proposed for partial constraint satisfaction (see [14] and references therein for some examples). Another variation of CSP aimed at relaxing hard constraints is Fuzzy CSP (FCSP) [15]. In FCSP, for each constraint, levels of preference between 0 and 1 are assigned to each variable assignment tuple. These levels indicate how "well" the assignment satisfies the constraint. A solution to FCSP is an assignment to all variables that maximizes the preference level of a constraint having the lowest satisfaction value. Yet another CP extension that should be mentioned in this context is the Probabilistic CSP [16]. In this framework, each constraint has an associated probability of presence in a "real" CSP. This framework allows to model the uncertainty regarding the presence of a constraint in a given problem. A solution to a probabilistic CSP is an assignment of values to variables that has the maximal probability to be a solution to the real problem.

A CSP formulation of a problem is processed by a constraint solver, which attempts to find a solution using a search algorithm combined with reductions of variable domains based on constraint information. The latter mechanism is known as *constraint propagation*.

The *search space* is the Cartesian product of all the variable domains. Let Z be a search space. A *solution space* $S \subseteq Z$ is a set of all possible assignments to variables that are solutions to the CSP.

Linear programming is a problem of finding an assignment to real (floating point) variables subject to linear constraints such that some linear function of these variables is minimized or maximized. Formally, a linear program can be expressed as

$$min \ cx \ subject \ to \ Ax \geq b, x \geq 0$$

where A is a matrix, c is a row vector, b is a column vector, and x is a vector of variables. There exist many additional equivalent forms of formulating an LP problem. Many algorithms have been developed for LP solving, e.g. the Simplex method. Quadratic programming is an extension of LP where the objective function is quadratic. Solution algorithms have been developed for certain forms of quadratic objective functions.

3   Distribution Constraints

In this section we formally introduce the notion of a distribution constraint. We observe that unlike a traditional CSP constraint, a distribution constraint is applicable not to a single variable assignment, but rather to a collection of multiple assignment tuples. Following the motivation discussed in the introduction, the purpose of this new concept is to provide the capability to constrain a collection of CSP solutions by specifying desired percentages of solutions with certain properties.

Definition 3.1: Given a set V of CSP variables, an *unconditional distribution constraint d* is defined by a pair *(c,p)*, where c is a CSP constraint on a subset of V and p is a real number such that $0 \leq p \leq 1$. A non empty collection S of assignments to V is said to *satisfy* an unconditional distribution constraint *d(c,p)* with precision error ε if *abs(N(S,c)//S| p)/p* = ε where $N(S,c)$ denotes the number of assignment tuples in S satisfying the constraint c.

In other words, an unconditional distribution constraint *d(c,p)* assigns a weight p to a subset of variable assignments which is given as a collection of tuples satisfying the constraint c. The precision error ε of satisfying d is the absolute difference between the actual percentage of assignments satisfying c in the collection S and the required percentage p, divided by p to get an error measure relative to the actual value of p.

For example, consider the following variables with their corresponding domains:

$$A[1..5], B[0..6], C[3..10]$$

Let an unconditional distribution constraint d be given by a constraint $c = A > B$ and $p = 0.6$. Then the collection S of 10 assignment tuples to (A, B, C) shown below satisfies d with the precision error $0.1/0.6 = 1/6$.

$$\{ (2,5,3),(4,0,10),(5,0,7),(3,6,9),(3,0,7),(1,4,3),(5,2,6),(2,1,10),(4,3,5),(3,1,4) \}$$

Indeed, 7 out of 10 tuples in S satisfy $A > B$, resulting in the actual percentage of 0.7 compared to the required percentage of 0.6.

Next, to enhance expressiveness in specifying the distribution requirements, we define an additional kind of distribution constraints, namely, a conditional distribution constraint. In fact, the latter is a generalization of an unconditional distribution constraint defined above. A conditional distribution constraint implies a distribution requirement not on all the assignment tuples in a given collection, but only on those satisfying a given condition. For example, in the context of functional test generation, a validation engineer might wish to test different modes of ADD instruction in interaction with other instructions. An example requirement in this case might be that a half of generated ADD instructions should have a memory operand and the other half should not. This requirement, however, should not affect any instructions whose mnemonic is not ADD. The formal definition of a conditional distribution constraint follows.

Definition 3.2: Given a set V of CSP variables, a *conditional distribution constraint* d_{cond} is defined by a 3-tuple *(cond,c,p)*, where *cond* and c are CSP constraints on a subset of V and p is a real number such that $0 \leq p \leq 1$. A collection S of assignments to V is said to *satisfy* a conditional distribution constraint $d_{cond}(cond,c,p)$ with precision error ε if $abs(N(S,cond \wedge c)/N(S,cond)$ $p)/p = \varepsilon$ where $N(S,constr)$ denotes the number of assignment tuples in S satisfying the constraint *constr*. We assume that $N(S,cond) > 0$, otherwise $d_{cond}(cond,c,p)$ is trivially satisfied.

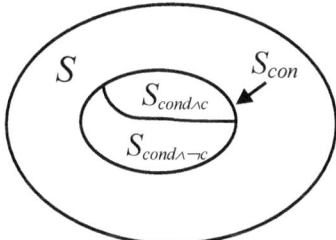

Fig. 1. Search space partition by a conditional distribution constraint

The partition of the solution space by a conditional distribution constraint is visualized in Fig.1. S_{cond} denotes a subset of tuples in S that satisfies the condition *cond*. This subset is divided into the subset $S_{cond \wedge c}$ of tuples that satisfy a constraint c and the subset $S_{cond \wedge \neg c}$ containing tuples that do not satisfy c. The conditional distribution constraint specifies the percentage p of tuples from $S_{cond \wedge c}$ within S_{cond}.

It can be easily seen that an unconditional distribution constraint introduced in Definition 3.1 is a special case of a conditional distribution constraint defined above with a condition *cond* equal to the constant true constraint. We will use the term *distribution constraints* to denote both conditional and unconditional distribution constraints. We will refer to a constraint c in the definition of distribution constraints as a *property*, to emphasize that this constraint specifies a variable assignment property to which the distribution requirement is implied.

We observe that definitions presented above differ from those in CP extension frameworks described in Section 2, where constraints are also associated with weights or probabilities. The principal difference is that in all of those frameworks weighted constraints are applied to a single variable assignment tuple and weights associated with constraints or with specific variable assignments are used to determine the best single solution. On the other hand, distribution constraints introduced above are applied to a collection of variable assignment tuples and are used in a problem of finding multiple solutions.

4   Problem Definition

We proceed with defining two versions of the problem of CP with distribution constraints. Both versions arise in the functional test generation. Formulations of both problems require distribution constraint set satisfaction by a collection of variable assignment tuples. Satisfaction of a distribution constraint set can be defined in multiple ways. Some of possible definitions are discussed in the sequel of this section.

The first problem, which we will refer to as Single CSP with Distribution Constraints (SCSPD), is to generate multiple solutions to the same CSP problem subject to distribution constraints. Formally, SCSPD is given by a variable set V, a superset Ω

of constraints over V, a CSP $P(V,C \subseteq \Omega)$, and a set of distribution constraints $D=\{(cond_j,c_j,p_j)\}_{1 \leq j \leq m}$ where $cond_j \in \Omega$, $c_j \in \Omega$. The objective is to find a collection S of assignments to V such that each $s \in S$ satisfies constraints in C and such that the distribution constraints set D is satisfied by S.

In functional test generation, the SCSPD problem can arise, for example, when it is required to generate multiple versions of the same instruction to test different modes of its operation. In this case, the CSP part of the problem would include architectural constraints implying the validity of the instruction, as well as constraints of the specific test scenario describing in particular which specific instruction needs to be tested. Distribution constraints in this application may require a specific occurrence ratio of different modes of the instruction in the sample of CSP solutions. Another example is the problem of generating multiple random tests from a single test specification, when distribution requirements apply to instructions with the same ordinal number in each test. For example, for a given test specification with unconstrained mnemonic of the first instruction, one would like to get 0.5 ratio of tests with the first instruction having mnemonic ADD, 0.3 ratio of tests with the first instruction SUB and 0.2 ratio of tests starting with MULT.

Next we define the second version of CP with distribution constraints which we call Multiple CSP with Distribution Constraints (MCSPD). In this problem a sequence of different CSPs sharing common variables is being solved, while distribution constraints apply to the sequence of these CSP solutions. Formally, MCSPD is given by a variable set V, a superset Ω of constraints over V, a sequence of CSP problems sharing a variable set V, namely, $P_1(V,C_1 \subseteq \Omega)$, $P_2(V,C_2 \subseteq \Omega)$, ..., $P_n(V,C_n \subseteq \Omega)$ where C_i denotes a constraint set of a problem P_i for $1 \leq i \leq n$, and a set of distribution constraints $D=\{(cond_j,c_j,p_j)\}_{1 \leq j \leq m}$ where $cond_j \in \Omega$, $c_j \in \Omega$. The objective is to find a sequence S of assignments to V such that for each $1 \leq i \leq n$, $s_i \in S$ satisfies constraints in C_i and such that the distribution constraints set D is satisfied by S. Observe that though each variable assignment in the sequence S is a solution to a different CSP problem, the variable set is common to all the tuples, which makes distribution constraints on these variables applicable to the whole sequence. We also note that ordering the collection S does not affect the definition of distribution constraint satisfaction by S.

An example of MCSPD in functional test generation is the problem of generating a stream of (different) test instructions. The problem of generating each instruction in the stream is described by an individual CSP with its own constraints, and distribution constraints specify the required ratios of instructions with specific properties, e.g. one can require that 0.7 ratio of instructions in the stream should be ADD.

Next we discuss the possible criteria for defining whether a collection S of CSP solutions satisfies a given set D of distribution constraints. In the previous section we defined the precision error of satisfying a distribution constraint d by a collection S of variable assignments. A fair criterion for satisfying a set of distribution constraints D by a given collection S of variable assignments should take into account precision errors $\varepsilon_d(S)$ of satisfying each $d \in D$ by S. The strict definition of satisfaction of D by S requires $\varepsilon_d(S)=0$ for each $d \in D$. However, using this strict definition makes many problem instances infeasible. Instead, one can use a relaxed definition of the satisfaction of D by S allowing non zero precision errors yet trying to keep these errors small.

There are multiple ways to simultaneously account for the precision errors $\{\varepsilon_d(S)\}_{d\in D}$. For example, if the distribution precision is of different importance for each distribution constraint, one can define a vector of upper bounds B_d on $\varepsilon_d(S)$ for each $d\in D$ and consider the set D satisfied by S if $\varepsilon_d(S)\leq B_d$ for each $d\in D$. On the other hand, if precision is equally important for all constraints one can define a common upper bound B on $\varepsilon_d(S)$ and require the precision error of any $d\in D$ not to exceed B. Another approach is to integrate all the precision errors $\varepsilon_d(S)$ into a single minimization criterion and define that D is satisfied by S if this criterion is minimized. A possible minimization criterion in this approach can be $\varepsilon_{avrg}=(\sum_{d\in D}\varepsilon_d)/|D|$. With this optimization criterion, the average precision error over $d\in D$ is minimized. Another possible choice for minimization criterion is $\varepsilon_{max}=max_{d\in D}\ \varepsilon_d$.

Another approach to distribution constraint satisfaction is to assume the strict definition with zero precision errors, but to look for approximate solutions to the problem. In this approach, the quality of the approximation can be measured using the same optimization criteria discussed above, e.g. ε_{avrg} or ε_{max}. This is the approach we take in the rest of this paper, with ε_{avrg} used as a measure of the approximation quality of the distribution constraint satisfaction.

Finally, recall that in the task of functional test generation there is an implicit requirement of generating representative solution samples. This requirement remains valid in the presence of distribution constraints, when SCSPD or MCSPD are solved in the functional test generation context. Therefore, samples created by replication of identical solutions in the required proportions could provide a theoretical solution but would be of little value in the functional test generation task.

5   Algorithms for CP with Distribution Constraints

In this section we present algorithms for solving the SCSPD and MCSPD problems defined in the previous section. We start with some preliminary discussion required for the presentation of our algorithms.

5.1   CSP Formulation for SCSPD and MCSPD

We observe that both SCSPD and MCSPD can be formulated as classical CSP problems. Without the loss of generality, we shall describe such formulation for MCSPD. SCSPD can be formulated in the same way as an MCSPD with a sequence of identical CSP problems. In the formulation below, we use the relaxed definition of distribution constraint satisfaction by means of the upper bound B on precision errors. Such relaxation will be required in the sequel for presentation of one of our algorithms. For the strict definition, one can substitute $B=0$.

Suppose we are given an instance of MCSPD defined by a variable set V, a superset of constraints Ω, a sequence of CSP problems $P_1(V,C_1),P_2(V,C_2),....,P_n(V,C_n)$, a set of distribution constraints $D=\{(cond_1,c_1,p_1),(cond_2,c_2,p_2),....,(cond_m,c_m,p_m)\}$, and an upper bound B on the precision errors of distribution constraint satisfaction. We denote the CSP corresponding to the MCSPD problem by $P_M(V_M,C_M)$, and define it as follows:

- $V_M{}^1 = V_1 \cup V_2 \cup \ldots \cup V_n$ where V_i is a replica of V, for $1 \le i \le n$;
- $C_M{}^1 = C_{1,1} \cup C_{2,2} \cup \ldots \cup C_{n,n}$ where $C_{i,i}$ is a replica of C_i defined over the variables in the replica V_i;
- $V_M{}^2 = \{b_{i,j}\}_{1 \le i \le n, 1 \le j \le m}$ where $b_{i,j}$ is an integer variable with domain $\{0,1\}$;
- $C_M{}^2 = \{b_{i,j}=1 \text{ iff } cond_{ij}\}_{1 \le i \le n, 1 \le j \le m}$ where $cond_{ij}$ denotes a replica of the distribution constraint condition $cond_j$ defined over V_i; that is, a variable $b_{i,j}$ equals 1 if and only if the condition $cond_j$ of the distribution constraint $(cond_j, c_j, p_j)$ holds over the replica V_i of variables;
- $V_M{}^3 = \{q_{i,j}\}_{1 \le i \le n, 1 \le j \le m}$ where $q_{i,j}$ is an integer variable with domain $\{0,1\}$;
- $C_M{}^3 = \{q_{i,j}=1 \text{ iff } cond_{ij} \wedge c_{ij}\}_{1 \le i \le n, 1 \le j \le m}$ where $cond_{ij}$ denotes a replica of the distribution constraint condition $cond_j$ and c_{ij} denotes a replica of the distribution constraint property c_j defined over V_i; that is, a variable $q_{i,j}$ equals 1 if and only if the condition $cond_j$ and the property c_j of the distribution constraint $(cond_j, c_j, p_j)$ holds over the replica V_i of variables;
- $C_M{}^4 = \{abs(\sum_{1 \le i \le n} q_{i,j} - p_j \sum_{1 \le i \le n} b_{i,j}) \le B \cdot p_j \sum_{1 \le i \le n} b_{i,j} \}_{1 \le j \le m}$; these constraints require the precision error for each distribution constraint not to exceed B;
- $V_M = V_M{}^1 \cup V_M{}^2 \cup V_M{}^3$
- $C_M = C_M{}^1 \cup C_M{}^2 \cup C_M{}^3 \cup C_M{}^4$

We observe that since $P_M(V_M,C_M)$ is formulated as a single problem of finding n solutions to $P_1(V,C_1),P_2(V,C_2),\ldots,P_n(V,C_n)$, the variable set V is replicated n times and the constraint C_i of each problem P_i are formulated on its corresponding replica of V. We apply the counting principle to express the number of solutions that satisfy the condition $cond_j$, for each of the given distribution constraints $(cond_j,c_j,p_j)$, as well as the number of the solutions satisfying $cond_j$ that also satisfy the property c_j. The former is expressed by $\sum_{1 \le i \le n} b_{i,j}$, while the latter is expressed by $\sum_{1 \le i \le n} q_{i,j}$. Finally, by definition, precision error ε for a distribution constraint $(cond_j,c_j,p_j)$ is expressed by $abs(N(S,cond \wedge c)/N(S,cond) - p)/p$, and the constraints $C_M{}^4$ guarantee that the precision error of any distribution constraint does not exceed the bound B. Randomization of solutions required in the context of functional test generation can be achieved, for example, by randomizing the value selection for variables by a CP search algorithm.

Observe that a solution of the CSP above with $B=0$ achieves an optimal solution to MCSPD in terms of the distribution precision error. However, for large CSP instances and large sequence lengths n (which is typically the case when one wishes to measure distribution) in the MCSPD definition, the resulting CSP problem P_M can be very large, and the CP search algorithm on such instance can require impractically high computational effort. Our experiments show that while CP search can be tuned to substantially improve performance on specific instances, this tuning is not universal but depends on the structure of a particular instance. Therefore, while the pure CP based approach provides high quality solutions on small instances, it suffers from the scalability problem and cannot be readily applied to large instances.

In Section 5.3 we show how the CP based approach can be used within a parallelization scheme to resolve the scalability problem while maintaining good precision quality. However, the latter approach is applicable only to the SCSPD version of the problem.

Before we can proceed to the presentation of the rest of the algorithms described further in this section, we need to introduce the concept of *feasible property subsets* with respect to a given set D of distribution constraints.

5.2 Feasible Property Subsets

For the clarity of presentation we start the introduction of the property subset concept for the unconditional distribution constraints, and then generalize it to encompass conditional distribution constraints. We observe that with respect to a distribution constraint $d(c,p)$, the solution space of a CSP can be divided into tuples that satisfy the property c and those that do not. Generally, with respect to a distribution constraint set $D=\{d_1(c_1,p_1),d_2(c_2,p_2),\ldots,d_m(c_m,p_m)\}$, a variable assignment s can be associated with a property subset $\lambda \subseteq \{c_1,c_2,\ldots,c_m\}$ so that $c_i \in \lambda$ if and only if s satisfies c_i, for $1 \leq j \leq m$. The set D of distribution constraints partitions the solution space into subspaces of tuples associated with the same property subsets. In each subspace, all the tuples of the subspace satisfy exactly the same subset of properties $\{c_1,c_2,\ldots,c_m\}$ and tuples from different subspaces satisfy different subsets of properties.

Consider the following example. Let $P(V,C)$ be a CSP instance with V containing two integer variables, $A[0..2]$ and $B[1..2]$, and $C=\{A \neq 2\}$. Let D be a set of two distribution constraints over V, $d_1(A=1, 0.3)$ and $d_2(A \geq B, 0.5)$. The partition of the solution space induced by these distribution constraints is shown in Fig. 2.

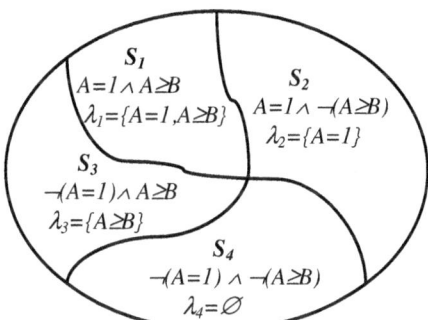

Fig. 2. Solution space partition by the unconditional distribution constraint set

Observe that in the example above the solution space contains the following assignment tuples to (A,B): $\{(0,1), (0,2), (1,1), (1,2)\}$. The distribution constraint set $\{d_1,d_2\}$ partitions this solution space into the following regions: $S_1=\{(1,1)\}$, $S_2=\{(1,2)\}$, $S_3=\varnothing$ and $S_4=\{(0,1),(0,2)\}$. Generally, some property subsets are feasible, meaning that the corresponding subspace of tuples satisfying the property subset is non-empty, and the other property subsets are infeasible, meaning that there are no tuples satisfying them.

Formally, given a CSP instance $P(V,C)$ and a set $D=\{d_1(c_1,p_1),d_2(c_2,p_2),\ldots,d_m(c_m,p_m)\}$ of distribution constraints, we call a subset of properties $\lambda \subseteq \{c_1,c_2,\ldots,c_m\}$ feasible if there exists a variable assignment that satisfies constraints in C and properties in λ and does not satisfy properties in $\{c_1,c_2,\ldots,c_m\}$ \ λ.

The definition of the solution space partition by a set of unconditional distribution constraints given above can be extended to accommodate conditional distribution constraints. The distinction of a conditional distribution constraint is in the fact that it divides the solution space into three regions and not into two regions like an unconditional distribution constraint. Specifically, a conditional distribution constraint $d(cond,c,p)$ partitions the solution space into tuples satisfying $cond \wedge c$, tuples satisfying $cond \wedge \neg c$ and tuples satisfying $\neg cond$, as shown in Fig. 1. Given a set of conditional distribution constraints $D=\{d_1(cond_1,c_1,p_1), d_2(cond_2,c_2,p_2), \ldots, d_m(cond_m,c_m,p_m)\}$, the regions of the solution space partition can be described as subsets of the extended property set including also conditions as its elements. If such subset λ contains both $cond_i$ and c_i then the corresponding tuples satisfy $cond_i \wedge c_i$, if λ contains $cond_i$ but not c_i, then the corresponding tuples satisfy $cond_i \wedge \neg c_i$, and if λ does not contain neither $cond_i$ nor c_i, then the corresponding tuples satisfy $\neg cond_i$. Since a region that does not satisfy $cond_i$ is not subdivided with respect to c_i there are no subsets containing c_i but not containing $cond_i$. For example, for a set of two conditional distribution constraints $D=\{d_1(cond_1,c_1,p_1), d_2(cond_2,c_2,p_2)\}$, the partition of the solution space by D can be described by the following property subsets (each property subset is followed by a constraint describing the corresponding region in the partition):

$$\varnothing \Rightarrow \neg cond_1 \wedge \neg cond_2$$
$$\{cond_1\} \Rightarrow cond_1 \wedge \neg c_1 \wedge \neg cond_2$$
$$\{cond_1,c_1\} \Rightarrow cond_1 \wedge c_1 \wedge \neg cond_2$$
$$\{cond_2\} \Rightarrow cond_2 \wedge \neg c_2 \wedge \neg cond_1$$
$$\{cond_2,c_2\} \Rightarrow cond_2 \wedge c_2 \wedge \neg cond_1$$
$$\{cond_1,cond_2\} \Rightarrow cond_1 \wedge \neg c_1 \wedge cond_2 \wedge \neg c_2$$
$$\{cond_1,c_1,cond_2\} \Rightarrow cond_1 \wedge c_1 \wedge cond_2 \wedge \neg c_2$$
$$\{cond_1,cond_2,c_2\} \Rightarrow cond_1 \wedge \neg c_1 \wedge cond_2 \wedge c_2$$
$$\{cond_1,c_1,cond_2,c_2\} \Rightarrow cond_1 \wedge c_1 \wedge cond_2 \wedge c_2$$

The approach taken in the algorithms presented further in this section is to generate the required number of solution samples from different subspaces in the partition described above. To guarantee that a CSP solution will belong to the subspace of the required property subset, properties from this subset are added to the CSP model along with the negations of properties that do not belong to the subset. The proportion of generated samples from each subspace should be chosen in a way satisfying the given distribution constraints.

In the approach described above, one needs to resolve two problems. The first one is to determine which property subsets are feasible, since adding model constraints corresponding to an infeasible property subset will result in an infeasible CSP instance. Another problem is to determine fractions of samples to generate for each feasible property subset so that distribution constraints are satisfied. The algorithms described in the rest of this section propose different ways of resolving these two problems.

5.3 CP Search Based Algorithm with Parallelization Scheme for SCSPD

In this subsection we apply the CP based approach described in Section 5.1 to determine both the feasible property subsets and the fractions of samples to be generated for each of these subsets. At the same time the proposed algorithm avoids scalability problems of the pure CP based approach by solving CSP on a small sequence.

We proceed with a formal presentation of the algorithm. The input of the algorithm is an instance of SCSPD, composed of a CSP instance $P(V,C)$, a set of distribution constraints $D=\{(cond_1,c_1,p_1),(cond_2,c_2,p_2),\ldots,(cond_m,c_m,p_m)\}$ and a number n of the required solution tuples. The algorithm proceeds as follows. At first, a CP based approach described in Section 5.1 is applied to build a small number k of samples, for some parameter $k<<n$. That is, a new CSP for obtaining a collection of k solutions of P subject to distribution constraints D is formulated as described in Section 5.1 and solved using a CP search algorithm. If the problem has no solution for $B=0$, a binary search on values of B can be applied to determine the smallest value of B for which the problem has a solution. Alternatively, the CSP described in Section 5.1 can be reformulated as an optimization problem for ε_{max}. When a collection of k solutions $S=\{s_1,s_2,\ldots,s_k\}$ is obtained, it is processed as follows. A partition of S with respect to property subsets is performed as described in Section 5.2. Clearly, all the property subsets that got a non empty representation in the partition are feasible, though there may also exist additional feasible property subsets that got an empty representation in the partition. Let $\lambda_1,\lambda_2,\ldots,\lambda_t$ be all the feasible property subsets that got non empty representation in the partition of S, and let $\eta_i>0$ be the number of samples in S corresponding to the property subset λ_i, for $1\leq i\leq t$. We observe that the actual ratio of samples satisfying a property or a condition c in S is given by $(\sum_{\{i|c\in\lambda_i\}}\eta_i)/k$, for $1\leq j\leq m$. We propose to use the same ratio η_i/k for each property subset λ_i in a solution collection of the full size n. From the expression above it follows that the ratio of samples satisfying c_j in a solution set of size n will be the same as that in the solution set S of size k, for $1\leq j\leq m$. Following the definition of precision error of distribution constraint satisfaction, this means that the precision error on the large solution collection will be equal to the precision error of the small collection.

Specifically, in the second phase of the algorithm, our objective is to build a collection of solutions to P of size n. Let n' be the smallest number such that $n'\geq n$ and n' is a multiple of k. We generate n' solutions from the solution subspaces corresponding to $\lambda_1,\lambda_2,\ldots,\lambda_t$ in the same proportions as in the "small" solution sample S of size k. To obtain a solution corresponding to λ_i, we add properties in λ_i and negations of properties not in λ_i as model constraints, as described in Section 5.2, and solve the resulting CSP using a randomized search algorithm to ensure diversity of solutions within the subspace corresponding to λ_i. The quantity of samples generated for the property subset λ_i is given by $(\eta_i/k)\cdot n'$. If the size of the solution sample must equal n exactly, we can discard n' $n<k$ of the generated solutions at random. Provided $k<<n$, this will introduce only a small distortion to the existing distribution.

We observe that the algorithm presented above resolves the scalability problem of the pure CP approach described in Section 5.1 since the CP search is performed on a smaller CSP problem. Moreover, the new algorithm can be effectively parallelized since the generation of solutions at the second phase of the algorithm can be divided into independent tasks. Specifically, a parallel thread can be assigned for each feasible property subset λ_i. A thread assigned for λ_i is responsible for generating all solutions corresponding to the property subset λ_i in the quantity of $(\eta_i/k)\cdot n'$. Since a thread calculates a number of solutions to exactly the same CSP problem, the search can be made more efficient relative to the search for a number of solutions to different CSP problems.

On the other hand, the precision error of distribution constraint satisfaction in the algorithm proposed above may be high in certain cases. Precision error occurs due to two factors. First, there might not exist appropriate $k<<n$ dividing the required sample size n without remainder. Then some solutions will be discarded at the final step of the algorithm distorting the distribution. Moreover, even if n is a multiple of k, the resulting precision error ε of satisfying D is optimal for a solution set of cardinality k, but not necessarily optimal for a solution set of cardinality n as the optimum precision error for a larger set can be less than that for a smaller set.

The algorithms described further in this section follow a completely different approach to finding feasible property subsets and calculating fractions of solutions to be generated for each feasible property subset. The description of these algorithms follows in the next two subsections.

5.4 Probabilistic LP Based Algorithm for SCSPD

The algorithm described in Section 5.3 assumed a deterministic approach to satisfying distribution constraints in the sense that ratios of solutions satisfying each property were determined by the algorithm deterministically. The next two algorithms we present assume probabilistic approach, and actual percentages of solutions satisfying each property are random numbers which are analyzed with respect to their expected values. Both algorithms are hybrid methods applying both CP and LP technology.

In this subsection we present a probabilistic algorithm for the SCSPD problem. Let an instance of SCSPD be composed of a CSP instance $P(V,C)$, a set of distribution constraints $D=\{(cond_1,c_1,p_1),(cond_2,c_2,p_2),\ldots,(cond_m,c_m,p_m)\}$ and a number n of required solution samples. We start with finding the feasible property subsets. The straightforward approach is to enumerate all the possible property subsets. For each property subset λ, corresponding constraints are added to the constraints in C, and the resulting CSP is solved by a CP search algorithm. If a solution was found, the corresponding property subset is marked feasible. The proposed approach is exponential in the number of distribution constraints m. However, in typical applications arising in the functional test generation, the number of simultaneous distribution requirements is small and the proposed method is practical. To improve the complexity of finding feasible property subsets by reducing the number of CP search invocations, a dynamic programming method can be applied. Specifically, let $\Psi=\{\varphi_1,\varphi_2,\ldots,\varphi_{2m}\}$ be an extended property set composed of the conditions $\{cond_1,cond_2,\ldots,cond_m\}$ and the properties $\{c_1,c_1,\ldots,c_m\}$. The improved algorithm for calculating the set of feasible subsets of Ψ proceeds in iterations so that at each iteration k a set Φ_k of all feasible subsets of properties $\{\varphi_1,\varphi_2,\ldots,\varphi_k\}$ is calculated, for $0\leq k\leq 2m$. Clearly, for $k=0$, $\Phi_0=\varnothing$. Given Φ_k, Φ_{k+1} can be computed by examining for each $\lambda\in\Phi_k$ two subsets of $\{\varphi_1,\varphi_2,\ldots,\varphi_k, \varphi_{k+1}\}$, namely, λ and $\lambda\cup\{\varphi_{k+1}\}$. For each of these two subsets, corresponding constraints are added to P and the resulting CSP problem is solved to determine the feasibility of the subset. If the subset proves feasible, it is added to Φ_{k+1}. Observe that if φ_{k+1} is a property and its corresponding condition does not belong to λ, then $\lambda \cup \{\varphi_{k+1}\}$ needs not be added to Φ_{k+1}. In the iterative approach described above, the number of CP search invocations is the order of the number of the *feasible* property subsets, compared to the number of *all* the property subsets in the straightforward enumeration.

Another way to speed up the calculation of feasible property subsets is to use parallel CSP solvers to check feasibility of each property subset instead of applying the dynamic programming scheme.

After the set of feasible property subsets is calculated, the algorithm proceeds to its second phase where it calculates the probabilities for generating a solution satisfying a feasible property subset λ, for each $\lambda \in \Phi_{2m}$. The objective is to calculate these probabilities in such a way that the expected precision error of satisfying each distribution constraint will be zero. This will be achieved if for each distribution constraint $(cond_i, c_i, p_i)$, for $1 \leq i \leq m$, the probability of generating a solution satisfying $cond_i \wedge c_i$ equals p_i multiplied by the probability of generating a solution satisfying $cond_i$. To calculate probabilities satisfying the required properties, we formulate and solve a linear program corresponding to the described problem. The linear program below is formulated as a set of linear constraints C_L over real variables V_L subject to linear minimization function O_L. We observe that such formulation can be transformed into an equivalent LP program in the form *min cx* subject to $Ax \geq b$, $x \geq 0$. The linear program $L(V_L, C_L, O_L)$ is defined as follows.

- $V_L = \{\rho_1, \rho_2, ..., \rho_t\}$ where $t = |\Phi_{2m}|$, $0 \leq \rho_j \leq 1$ for $1 \leq j \leq t$. Each variable ρ_j represents the probability of generating a solution satisfying the feasible property subset $\lambda_j \in \Phi_{2m}$
- $C_L^{\ 1} = \{\sum_{1 \leq j \leq t} \rho_j = 1\}$. This constraint follows from the fact that Φ_{2m} defines a partition of the solution space
- $C_L^2 = \left\{ \sum_{\{ j | cond_i \in \lambda_j \wedge c_i \in \lambda_j \}} \rho_j - p_i \sum_{\{ k | cond_i \in \lambda_k \}} \rho_k = 0 \right\}_{1 \leq i \leq m}$
- $C_L = C_L^{\ 1} \cup C_L^2$
- $O_L = const$

Note that the definition above does not require optimization of feasible solutions. However, to improve the diversity of generated solutions for SCSPD, one can apply a quadratic minimization objective $\sum_{1 \leq j \leq t} (\rho_j - \rho_{avrg})^2$ where ρ_{avrg} stands for the average probability $(\sum_{1 \leq j \leq t} \rho_j)/t$. The effect of such optimization criterion would be a more equal distribution of probability weights between different feasible property subsets.

After the probabilities for each feasible property subset are calculated, the algorithm generates solutions to $P(V,C)$ so that a solution satisfies a property subset λ_j with probability ρ_j, for $1 \leq j \leq t$. This is done by random selection of a subset λ_j out of Φ_{2m} with probability ρ_j and adding the constraints corresponding to λ_j (as described in Section 5.2) to the CSP model of P. The augmented model is then solved by a randomized CP search algorithm to produce a random solution satisfying λ_j.

It is easy to see that the proposed algorithm is an optimal probabilistic algorithm for SCSPD as it achieves a zero expected distribution precision error. It also has an advantage in terms of diversity over the algorithm presented in Section 5.3 since it allows sampling of diverse regions of solution space partition with respect to property subsets. A disadvantage of the probabilistic approach is that on small sample sizes there can be a substantial deviation from the expected precision error value.

5.5 Probabilistic LP Based Algorithm for MCSPD

The idea of LP based algorithm for MCSPD is similar to that presented in the previous subsection. However, while the previous algorithm calculated probabilities for feasible property subsets once and used them in creating all the solution samples, the same method cannot be used for MCSPD, where feasible property subsets can change for each CSP problem in MCSPD sequence. To overcome this problem, we calculate feasible property subsets for each CSP in the sequence and solve a large LP program that calculates the probability for each feasible property subset in each CSP instance in the sequence. The formal presentation of the algorithm follows below.

Let an instance of MCSPD be given by a sequence of CSP problems $P_1(V,C_1),P_2(V,C_2),...,P_n(V,C_n)$ and a set of distribution constraints $D=\{(cond_1,c_1,p_1),(cond_2,c_2,p_2),...,(cond_m,c_m,p_m)\}$ over V. The algorithm starts with calculating the sets of feasible property subsets for each of $P_1,P_2,...,P_n$. Let $\Phi_1,\Phi_2,...,\Phi_n$ denote the resulting sets, and let $t_i=|\Phi_i|$ for $1\leq i\leq n$. To reduce the number of CP search invocations in the calculation of these sets, we first calculate a set Φ of feasible property subsets in the absence of any model constraints and then obtain Φ_i from Φ by discarding those property subsets that become infeasible in the presence of constraints C_i, for $1\leq i\leq n$.

Once the sets of feasible property subsets are found, the algorithm calculates the probabilities ρ_{ij} of generating a solution satisfying a feasible property subset $\lambda_{ij}\in\Phi_i$ for the CSP problem P_i in the sequence, for $1\leq i\leq n$, $1\leq j\leq t_i$. These probabilities are calculated by formulating and solving the following linear program $L(V_L,C_L,O_L)$:

- $V_L=\{\rho_{ij}\}$ for $1\leq i\leq n$, $1\leq j\leq t_i$. Each variable ρ_{ij} represents the probability of generating a solution to P_i satisfying the feasible property subset $\lambda_{ij}\in\Phi_i$.

- $C_L^1 = \left\{\sum_{1\leq j\leq t_i} \rho_{ij} = 1\right\}_{1\leq i\leq n}$. These constraints follow from the fact that Φ_i defines a partition of the solution space of P_i.

- $C_L^2 = \left\{\sum_{1\leq k\leq n,\{i,j|cond_i\in\lambda_{kj}\wedge c_i\in\lambda_{kj}\}} \rho_{ij} - p_i\sum_{1\leq k\leq n,\{i,j|cond_i\in\lambda_{kj}\}} \rho_{ij} = 0\right\}_{1\leq i\leq m}$

- $C_L = C_L^1 \cup C_L^2$

- $O_L = const$

After the probabilities ρ_{ij} are calculated, the algorithm generates a solution to each P_i, for $1\leq i\leq n$, by randomly selecting a feasible property subset $\lambda_{ij}\in\Phi_i$ with probability ρ_{ij}, and adding the corresponding constraints to the CSP model of P_i to find a solution satisfying the property subset λ_{ij}.

We observe that like the algorithm for SCSPD described in the previous subsection, the algorithm described above also achieves zero expected precision error in distribution constraint set satisfaction.

An additional advantage of LP based algorithms for SCSPD and MCSPD is that they can be parallelized in their last stage of solution generation, since the generation of each solution in both algorithms is done independently.

Finally, we note that the usage of LP programming provides a powerful capability of solution failure explanations. In many unsolvable cases LP algorithms can provide information on minimal changes that can be done to bounds in linear constraints in order to make the problem instance feasible. This information is a valuable hint for a user in case of infeasible distribution constraints on how the distribution requirements could be modified in order to obtain a solution, as well as a debug aid in case of incorrect distribution constraint formulation.

6 Experimental Results

We have tested the algorithms described in Section 5 for different solution sample sizes. The goal of the experiments was to demonstrate and compare performance and distribution precision quality of the proposed algorithms. Tables 1 and 2 below summarize the results of our experiments. The experiments were performed on two generic CSP instances. The first instance included 10 distribution constraints with multiple conflicts within the superset of model constraints, distribution conditions and distribution properties providing a hard to satisfy test case. The second instance contained 4 distribution constraints with less conflicts; this instance was much easier to satisfy than the first one. For MCSPD, the sequence of different CSP problems was obtained by augmenting the CSP instances described above with some additional model constraints varying for each problem in the sequence. In our experiments we used ILOG CP as a CP engine and ILOG CPLEX as an LP engine [17]. For each algorithm, solution samples of sizes 100, 500 and 1000 were generated on 10 different random seeds. Entries of Table 1 show the average running time in seconds of the algorithms along with the average precision error ε_{avrg} of satisfying the given distribution constraint set. We performed the evaluation on Pentium M 2.5 GHz processor with 2.96 GB of RAM.

The results of our experiments confirm the previously mentioned theoretical conclusions regarding strengths and weaknesses of the proposed algorithms. Specifically, the results show that the CP based algorithm suffers from less accurate distribution precision, following from the fact that the frequencies of property subset occurrence are calculated on small size solution samples. On the other hand, this algorithm achieves better performance on computationally simple problem instances. The LP based algorithm for SCSPD achieves the best tradeoff between the running time and the distribution precision error. Its precision is less accurate on small solution samples where the deviation from the expected zero precision value is large, but the precision improves as the sample size is increased. Finally, as can be seen experimentally, the LP based algorithm for MCSPD achieves almost zero precision error. This is due to the fact that the LP algorithm tends to find an almost integral solution for probabilities of feasible property subsets. Clearly, a fully integral solution is an exact and therefore the best possible solution to the problem of finding solution with zero precision error. However, as the experiments show, this algorithm requires a larger running time. We also observe that this algorithm is the only one of the three proposed algorithms that can solve the MCSPD version of CP with distribution constraints. In addition, an important advantage of the LP based algorithms is that their performance is less sensitive to the complexity of the problem instance, as the counting is done not in a CSP, like in the CP based algorithm, but by means of LP, eliminating the need for a complicated CP search tuning.

Table 1. Experimental results for the sparsely and densely constrained test cases

	100 solutions		500 solutions		1000 solutions	
Test case	Sparse	Dense	Sparse	Dense	Sparse	Dense
CP based algorithm for SCSPD	0.001s 0.150	1.38s 0.200	0.015s 0.150	1.41s 0.200	0.016s 0.150	1.48s 0.200
LP based algorithm for SCSPD	0.027s 0.087	0.14s 0.150	0.091s 0.039	0.26s 0.083	0.15s 0.024	0.44sec 0.058
LP based algorithm for MCSPD	0.228s 0.015	1.39s 0.008	1.08s 0	6.80s 0.006	2.19s 0.001	13.40s 0

7 Conclusion

In this paper we addressed the novel concept of CP with distribution constraints motivated by CP applications arising in the functional test generation field. We presented the formal framework for this concept, including definitions of a distribution constraint and of two different versions of the problem of CP with distribution constraints, refining the previous attempts to formalize this problem. The new framework provides a well defined measure for the quality of distribution constraint satisfaction, providing a sound basis for comparison of different algorithms for this problem. We presented several algorithms to solve each of the problem versions, applying both deterministic and probabilistic approaches to distribution constraint satisfaction. The algorithms applying the deterministic approach provide a controllable tradeoff between the quality of distribution and the running time, ranging from an optimal distribution quality algorithm with high running time to a good approximation of the required distribution with a better running time. The algorithms applying the probabilistic approach achieve the optimal expected value of distribution quality measure. The proposed framework and solution methods extend the capabilities of automated functional test generation tools and address important requirements in the validation domain.

References

1. Bin, E., Emek, R., Shurek, G., Ziv, A.: Using a constraint satisfaction formulation and solution techniques for random test program generation. IBM Systems Journal 41(3), 386–402 (2002)
2. Naveh, Y., et al.: Constraint-based random stimuli generation for hardware verification. AI Magazine 28(3), 13–18 (2007)

3. Gutkovich, B., Moss, A.: CP with architectural state lookup for functional test generation. In: 11-th Annual IEEE International Workshop on High Level Design Validation and Test, pp. 111–118 (2006)

4. Moss, A.: Constraint patterns and search procedures for CP-based random test generation. In: Yorav, K. (ed.) HVC 2007. LNCS, vol. 4899, pp. 86–103. Springer, Heidelberg (2008)

5. Gomes, C., Sabharwal, A., Selman, B.: Near-uniform sampling of combinatorial spaces using XOR constraints. In: Schölkopf, B., et al. (eds.) Advances in Neural Information Processing Systems, vol. 19, pp. 481–488. MIT Press, Cambridge (2007)

6. Dechter, R., et al.: Generating random solutions for constraint satisfaction problems. In: 18-th National Conference on Artificial Intelligence, pp. 15–21 (2002)

7. Van Hentenryck, P., Coffrin, C., Gutkovich, B.: Constraint-based local search for the automatic generation of architectural tests. Accepted to 15-th International Conference on Principles and Practice of Constraint Programming (2009)

8. Larkin, D.: Generating solutions to constraint satisfaction problems with arbitrary random biasing. In: 8-th International Conference on Principles and Practice of Constraint Programming, Doctoral Programme (2002)

9. Moss, A.: Modeling for Random Search with Distribution-Oriented Constraints. In: 18-th European Conference on Artificial Intelligence, Workshop on Modeling and Solving Problems with Constraints, Patras, Greece, pp. 63–68 (2008)

10. Smith, B.M.: Modeling for constraint programming. In: The 1-st Constraint Programming Summer School (2005)

11. Rudova, H.: Constraint Satisfaction with Preferences. Ph.D. Thesis, Masaryk University, Brno (2001)

12. Schrijver, A.: Theory of Linear and Integer Programming. Wiley, Chichester (1998)

13. Freuder, E., Wallace, R.: Partial constraint satisfaction. Artificial Intelligence 58, 21–70 (1992)

14. Petit, T., Bessière, C., Régin, J.C.: A general conflict-set based framework for partial constraint satisfaction. In: 5-th International Workshop on Soft Constraints, Kinsale, Ireland (2003)

15. Dubois, D., Fargier, H., Prade, H.: The calculus of fuzzy restrictions as a basis for flexible constraint satisfaction. In: 2-nd IEEE International Conference on Fuzzy Systems, vol. 2, pp. 1131–1136 (1993)

16. Fargier, H., Lang, J.: Uncertainty in constraint satisfaction problems: A probabilistic approach. In: Moral, S., Kruse, R., Clarke, E. (eds.) ECSQARU 1993. LNCS, vol. 747, pp. 97–104. Springer, Heidelberg (1993)

17. ILOG Optimization Documentation (2007)

An Explanation-Based Constraint Debugger

Aaron Rich, Giora Alexandron, and Reuven Naveh

Cadence Design Systems
{ari,giora,rnaveh}@cadence.com

Abstract. Constraints have become a central feature of advanced simulation-based verification. Effective debugging of constraints is an ongoing challenge. In this paper we present GenDebugger - an innovative GUI-based tool for debugging constraints. GenDebugger shows how a set of constraints is solved in a step by step flow, allowing the user to easily pick out steps that are key to the buggy behavior. It is unique in that it is designed for dealing with various types of constraint-related problems, not only constraint-conflicts or performance issues. GenDebugger utilizes principles from the explanation-based debug paradigm [1], in that it displays information justifying each step of the solution. A unique contribution is that it enables the user to easily combine separate explanations of individual steps into a comprehensive full explanation of the constraint-solver's behavior. GenDebugger was lately released for full production, after an extensive process of evaluation with early access users. Our own experience and our users feedback indicate that GenDebugger provides highly valuable help in resolving problems found in constraints.

Keywords: Constrained-Random Verification, Constraint Debugging.

1 Introduction

Constrained Random generation of stimuli is a technique widely used in advanced simulation-based verification. To verify a Device Under Test (DUT) test scenarios are generated randomly but nevertheless must comply with constraints set by the verification engineer. This technique is supported by various verification-oriented languages such as *e* and System Verilog. Coding errors in constraints are revealed in several forms, such as unexpected values assigned to variables, unexpected distribution of values across a set of simulations, or failure to find a solution due to conflicting constraints. Bad runtime and memory consumption of constraint solving activity is also a common problem.

Debugging of constraints poses a considerable challenge (see [3], for example) because constraints reflect complex relationships between variables, and might depend on complicated conditions. What is more, regular debugging tools such as source-line debuggers, with which most software engineers are familiar, are ill-suited for debugging constraints. The reason is that source-line debuggers are normally sequential, showing and controlling the line-by-line imperative flow of user code. Constraints, in contrast, are declarative entities, analyzed and solved

K. Namjoshi, A. Zeller, and A. Ziv (Eds.): HVC 2009, LNCS 6405, pp. 52–56, 2011.

by a constraint solver built within the verification software. Because of this the processes governing constraint-solving are not visible to the user and this renders basic concepts of the common debugger paradigm, such as breakpoints and single-stepping, useless.

One approach to constraint debugging is to remove or add constraints until the generated stimuli is satisfactory. This is not necessarily a viable solution, given that it doesn't explain the problem with the original, preferred set of constraints. A second approach is to have the solver print the constraints directly involved in a constraint conflict, but this isn't appropriate for other types of problems. Even for constraint conflicts, it provides very limited help when a complex set of constraints is involved. The human mind can only deal with a limited amount of information at once. Another common debugging method is to trace the constraint solver's activity [2] to a file or the screen, but again - the multitude of static information a user must deal with makes this a very limited solution. A relatively new and promising direction is explanation-based debugging, where a debugger displays information justifying decisions of the constraint solver [1].

2 GenDebugger - Explanation-Based Debugging

In this paper we describe GenDebugger, the generation debugger of IntelliGen, the new generation engine of the testbench automation tool Specman. GenDebugger depicts the constraint-solving process as a sequence of steps, each reducing the domain of one or more variables, until each variable has a single value that complies with the constraints. Several types of steps are shown, including domain reductions based on constraints, value assignments, backtracking, input sampling, and more.

GenDebugger shows each step in a detailed, interactive view. The variables and constraints involved in the step are displayed, so that it can be clearly seen which variables were reduced and which constraints were involved. Because solving steps typically reduce the domains of variables, the post-step domains are displayed together with the pre-step domains, so that the nature of the change in a variable's domain is evident.

GenDebugger is interactive in that any variable or constraint can be selected and queried. A dedicated panel shows various pieces of information such as the type, path, or source of a selected variable, or the source and declaring class of a selected constraint. A specially important piece of information for a variable is its list of steps (see section 2.2).

2.1 GenDebugger and Explanations

GenDebugger explains solving-steps. That is, it displays information justifying the step, usually a subset of constraints and the domains of related variables. For example, a reduction of the domains of the unsigned integers x and y to

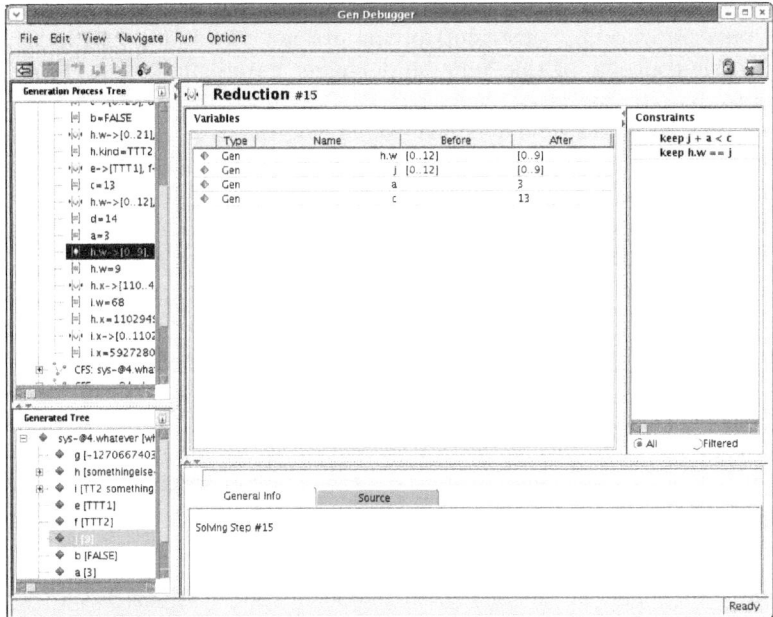

Fig. 1. Explaining a Reduction

[1..9] and [0..8] respectively, is justified by the constraints $x<10$ and $y<x$. After assigning 5 to y, a later step further reducing x's domain to [6..9] is explained by the constraint $y<x$ together with the value assigned to y.

Naturally, a solving step is easier to explain when it involves only a few bits of information. Because of this, when showing a solution to a large set of constraints, GenDebugger breaks the solution into several smaller steps, each much more explainable than the entire process as a whole. See figure 1 to see how GenDebugger explains a single reduction.

2.2 Full Explanations

It is common that some of the elements of a step's explanation are themselves solving steps, performed earlier in the solving process. For example, a reduction on z to [0..4], explained by the constraint $z<x-1$ coupled with x's domain: [6..9], raises the question: why was the domain of x [6..9] in the first place? The explanation is not complete without understanding the earlier reduction on x setting it's domain to [6..9].

GenDebugger is designed to enable the user to easily get a *full explanation* for a step. It does this by providing easy access to other, potentially relevant, steps. When any variable involved in the current step is selected, all steps affecting this variable are listed (see figure 2). A simple double-click displays the explanation for this second step. In this way the user combines explanations of several related steps into a full and comprehensive explanation. To continue our example, when viewing the reduction on z, selecting x lists all steps performed on x, including the reduction to $[6..9]$. Double-clicking on this earlier reduction, causes GenDebugger to display its explanation. The combination of these two explanations together form the full explanation for the reduction of z to $[0..4]$.

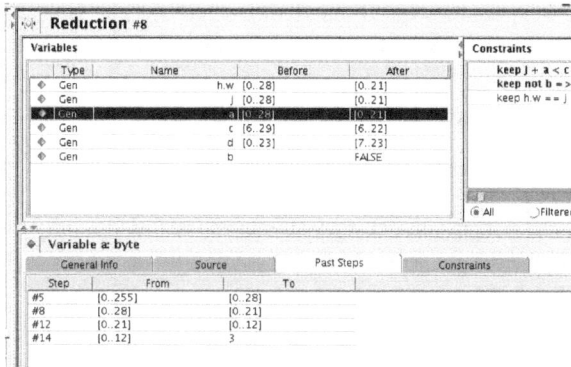

Fig. 2. All Steps on Variable a

3   Debugging Various Constraint-Related Problems

3.1   Understanding a Constraint Conflict

A constraint conflict is depicted by GenDebugger as a failed attempt to reduce variable domains due to contradicting constraints. When faced with a constraint conflict, the user sets a breakpoint telling GenDebugger to open when the conflict occurs (e.g., *break on gen error*). Then, when opened, GenDebugger automatically displays the failing step's explanation. The explanation can be explored further to form a full explanation, as described above in 2.2.

3.2   Understanding Value Assignments

Sometimes a variable is assigned a value that seems to the user unreasonable. In this case the user can set a second type of breakpoint telling GenDebugger to open when this variable's value is generated (e.g. *break on gen field packet.data*). It then opens to show how the variable is generated. Selecting the assignment-step from the list of steps on the variable, the user gets the required explanation. This can be continued to other, related steps until a full explanation of the unexpected assignment is formed.

Another way to have GenDebugger display the assignment to a variable, after it occurred, is via a simple command (e.g. *show gen x*). If information on the generation of the variable is accessible at that time, GenDebugger opens and automatically displays the explanation for the assignment.

3.3 Understanding Distribution

Distribution issues arise when, across many tests, the entire range of values generated for a variable, or the probability of certain values, is unexpected. To debug this type of issue, the user can utilize the methods described in the previous section 3.2. The user explores what determined the value assigned to the variable in a specific test. This exploration should reveal the constraints blocking, or restricting the probability of, some values from being assigned.

3.4 Understanding Performance

When the system seems stuck on a constraint-solving process, the user can break the test and open GenDebugger. GenDebugger opens on the solving step currently performed by the solver, and the user can also easily see all the steps preceding it. This is likely to reveal what is requiring so much solving activity, and why.

A common reason for performance issues is numerous backtracking. When this occurs, GenDebugger shows many backtrack steps in its list of solving steps. The explanation of a backtrack step is a bit different. It is made of the entire sequence of steps that were cancelled by the backtrack, and that terminated in a failure. GenDebugger shows the cancelled track of steps, allowing the user to understand why it ended at a dead end.

4 GenDebugger Experience

GenDebugger is the culmination of more than 10 years of experience in debugging constraint-solving issues. It was designed based on user feedback, with various types of verification needs. GenDebugger was lately released for full production after an extensive process of evaluation and validation with early access customers. The overall reaction to GenDebugger is no less than enthusiastic. According to our users, there is no comparison between GenDebugger and the earlier constraint-debugging solutions. We feel this is a promising avenue for significantly reducing the effort of creating constraint-based verification environments.

References

1. Ghoniem, M., Jussien, N., Fekete, J.D.: VISEXP: Visualizing Constraint Solver Dynamics Using Explanations. In: Seventeenth International Florida Artificial Intelligence Research Society Conference, FLAIRS 2004, Miami Beach, FL. AAAI press, Menlo Park (May 2004)
2. Meier, M.: Debugging Constraint Programs. In: Saraswat, V., Hentenryck, P.V. (eds.) Principles and Practice of Constraint Programming. LNCS, vol. 976, pp. 204–221. MIT, Cambridge (1995)
3. Simonis, H., Aggoun, A.: Search-Tree Visualisation. In: Deransart, P., Hermenegildo, M.V., Maluszynski, J. (eds.) DiSCiPl 1999. LNCS, vol. 1870, pp. 191–208. Springer, Heidelberg (2000)

Evaluating Workloads Using Multi-comparative Functional Coverage

Yoram Adler[1], Shmuel Ur[1], and Dale Blue[2]

[1] IBM Haifa Research Lab, University Campus, Carmel Mountains, Haifa, 31905, Israel
[2] IBM Systems & Technology Group, 2455 South Rd., Poughkeepsie, NY 12601, USA
adler@il.ibm.com, ur@il.ibm.com, dblue@us.ibm.com

Abstract. In this paper we present a technique for comparing multiple tests and workloads. We show how to automatically determine what each test does uniquely and how to present the information as succinctly as possible. This technology has a number of uses, including in understanding the contribution of tests in regression buckets, especially of legacy systems, and in evaluating what is missing in tests compared to customer usage. We also show that the technology used in the analysis is superior to previous technology in that it can automatically find holes that were previously only found manually.

Keywords: functional coverage, testing, regression.

1 Introduction

In software testing code coverage is a commonly known software testing technique [5]. Code coverage records the locations in the code that have been traversed during test execution. Many coverage tools can be used to guide the test effort [12] as well as to select tests [7] and prioritize them [6]. Code coverage, by its nature, is syntactic as it requires no specific understanding of the application. This is one of its strengths, as it is simple to apply and its feedback is easy to understand.

Functional coverage [8] [11] is a coverage methodology for evaluating the completeness of testing against application-specific coverage models. Once it became evident [8] that application-specific models still have many processes that can be automated, a number of functional coverage tools were created to handle all the common requirements. These tools include Meteor [9], Specman Elite [14], and FoCuS [15]. While functional coverage is in use in many companies [4] [13] and projects, it is traditionally more common to the world of hardware verification than to software testing.

From the outset, functional coverage tool development has concentrated on exploring data in a variety of ways [4], including the ability to view projections of subsets of the attributes or values. Another useful view is hole analysis, which automatically discovers large sets of uncovered tasks that have something in common [9].

The ability to explore data easily in a variety of ways is the main reason functional coverage tools are preferred to model-specific tools. It is very easy to conceive of a way to view the data and then write a script that processes and presents the data in this

K. Namjoshi, A. Zeller, and A. Ziv (Eds.): HVC 2009, LNCS 6405, pp. 57–70, 2011.
© Springer-Verlag Berlin Heidelberg 2011

specific way. However, when you collect the data, you do not know exactly how you will want to look at it. Furthermore, the process of exploring the data is interactive in nature. We estimate that during half an hour of exploring the data, we may examine more than ten different views. If we had to write a script for each view, the cost would quickly become prohibitive.

In *Evaluating Workloads Using Comparative Functional Coverage* [1] we showed how to enhance the exploration capability of functional coverage tools to evaluate the difference between workloads used to emulate customer activity and actual workloads collected from the customer. We started the exploration by working concurrently on two equivalent models with two separate sets of input data: one each from the customer and the client. While it is easy to manipulate each set of data, comparing them is not trivial. For example, it is easy to see how many tests for each model have been executed by the customer or in test, but it is not easy to see which of them comprises a larger portion of the relevant workload. For this reason we enhanced FoCuS, a functional coverage tool with the ability to look simultaneously at multiple data sets. Once this capability was in place we found many additional applications for the view. For example, presenting the code coverage of a test as one workload and changes from the version control as the second workload, and then sorting this view such that the tasks with the most changes compared to coverage are on top enables one to see code that was changed but is untested. Such untested code is at high risk of containing errors.

In this paper we show how to expand the technique to compare multiple coverage sources to one. In the comparison we utilize a new code coverage view for large applications called substring hole analysis [2], [3]. For code coverage the view is used for improving coverage and to define new tests cases. Substring hole analysis capitalizes on the fact that functions are given meaningful names; for example, "OpenDwarfSlowWithoutException". A substring hole is a string that is common to many functions, most of which have not been covered. For example, in [3] we show analysis of an application with more then one hundred thousand functions. One of the substring holes is marked as Dwarf 315 315, meaning that there are 315 functions containing the substring "Dwarf", none of which have been covered.

In this paper we show how comparing multiple coverage sources is useful for a number of applications. We explain first what it means to compare a number of coverage sources and how it is done and then show its usefulness on a number of applications. We also examine complicated cases where the straightforward approach fails, explaining how to deal with them. Our assumption is that, like the work on comparative functional coverage, once this view becomes available, additional uses will be found for it.

2   Algorithms and Tooling

This section presents the algorithms, tools, and methodology. Some of the issues we discuss are:

- Methods for comparing many traces
- An automatic normalization algorithm for traces that contain data from tests with different running times

- A method for naming the distinguishing features of each trace
- Different comparison modes, including grouping the traces

2.1 Fundamentals and Definitions

Applying coverage usually entails choosing a coverage model, collecting traces, and comparing the traces with the model. The coverage model is composed of a list of tasks, where each task is a binary function on a test that is true if the task has been performed by the test (for example, a line was executed). A task list is maintained, indicating for each task how many times it has been covered. Figure 1 shows an example of a coverage report. In this report, each task has five attributes: APIname, ServiceName, Duration, NumOther, and SameSrvcOnSystem. In the table, the highlighted line represents a task with a value of ATRCMIT for the APIname attribute, Commit_UR for the ServiceName attribute, medium for the Duration attribute, low for NumOther attribute, and 1 for the SameSrvcOnSystem attribute, which was covered 17228 times. This report was produced using the FoCuS tool.

Here are the legal tasks :

Help

Number of Tasks: 52560   Covered: 64   Coverage Percentage: 0.1%

Click on a column header to sort the rows by this column

APIname	ServiceName	Duration	NumOther	SameSrvcOnSystem	COUNT
ATRPDUE	Post_Deferred_UR_Exit	short	high	1	28417
ATRCMIT	Commit_UR	medium	low	1	17228
ATREINT5	Express_UR_Interest	short	low	0	16920
ATREINT5	Express_UR_Interest	short	high	0	14831
ATREINT5	Express_UR_Interest	short	0	0	7951
ATRSUSI2	Set_Side_Information	short	high	0	7702
ATRSUSI2	Set_Side_Information	short	high	1	6583
ATREINT	Express_UR_Interest	short	high	0	6252
ATREND	End_Transaction	short	high	0	5877
ATRCMIT	Commit_UR	medium	0	0	4682
ATRCMIT	Commit_UR	long	low	1	2161
ATRPDUE	Post_Deferred_UR_Exit	short	high	0	1568

Exclude   Export

Fig. 1. Coverage results for a single data source

2.2 Comparing Raw Data from Multiple Data Sources

When comparing multiple data sources (collecting traces) as shown in Figure 2 (in this case, six data sources are used), the coverage tasks list has measurements from each source. The coverage of trace [n] data is under the [n] column. In the view shown in the figure, we sort the data so that tasks with large changes between the sources are on top. Line 73 shows a task covered by only one data source. The figure is part of a prototype program that implements the multi-trace compare feature, which will soon be added to the FoCuS tool. The results from the data sources are sorted by ratio between one data source against all the others. Line 73 shows a task whose trace #2 has better coverage than all the other traces. The task is represented by a string in the right column. The string was constructed by concatenating the values of the five model attributes. Line 72 shows the task in which Trace #1 has the best coverage, while lines 73 to 78 show six tasks for which Trace #2 has better coverage than other traces.

68	Compare by tasks Ratio on Top:					
69						
70	1	2	3	4	5	6
71						
72	- 81 -	22	22	0	0	0	ATRCMIT-Commit_UR-long-0-0
73	0	- 1 -	0	0	0	0	ATRIBRS-Begin_Restart-long-low-1
74	0	- 1 -	0	0	0	0	ATRIERS-End_Restart-short-low-0
75	0	- 1 -	0	0	0	0	ATRIRLN-Retrieve_Log_Name-long-low-0
76	0	- 1 -	0	0	0	0	ATRIRLN-Retrieve_Log_Name-long-low-1
77	0	- 1 -	0	0	0	0	ATRISLN-Set_Log_Name-long-low-0
78	0	- 1 -	0	0	0	0	ATRISLN-Set_Log_Name-long-low-1
79	0	0	0	- 1 -	0	0	ATREND-End_Transaction-long-high-1
80	0	0	0	38	- 7270 -	0	ATRPDUE-Post_Deferred_UR_Exit-short-high-0
81	0	0	0	0	- 1 -	0	ATRIRLN-Retrieve_Log_Name-long-0-0
82	0	0	0	0	- 1 -	0	ATRISLN-Set_Log_Name-long-0-0

Fig. 2. Ratio comparison between six traces on a functional coverage model

Sorting the tasks using the "changed on top" view first displays tasks with a high absolute value of change between the sources. This sorting capability enables the user to find out where one workload focuses and another lacks focus. An example of this view, which is useful in the exploration stage, is shown in Figure 3. One can see that the task in Line 72, which is taken from trace #5, has the highest absolute value change compared to all other tasks.

68	Compare by tasks Changed on Top:					
69						
70	1	2	3	4	5	6
71						
72	0	0	0	38	- 7270 -	0	ATRPDUE-Post_Deferred_UR_Exit-short-high-0
73	0	0	0	0	0	- 31 -	ATREINT-Express_UR_Interest-medium-high-0
74	0	0	0	0	0	- 9 -	ATREND-End_Transaction-medium-high-1
75	0	0	0	0	0	- 9 -	ATRSUSI2-Set_Side_Information-short-medium-1
76	0	0	0	2	0	- 733 -	ATRSUSI2-Set_Side_Information-medium-high-1
77	0	0	0	0	0	- 7 -	ATREINT-Express_UR_Interest-short-0-0
78	0	0	0	0	0	- 5 -	ATREND-End_Transaction-short-0-0
79	0	0	0	0	0	- 4 -	ATRPDUE-Post_Deferred_UR_Exit-medium-medium-0
80	0	0	0	0	0	- 3 -	ATREND-End_Transaction-short-low-1
81	0	0	0	2	0	- 202 -	ATRPDUE-Post_Deferred_UR_Exit-medium-high-1

Fig. 3. Changed On Top comparison on a functional coverage model

In each of the above figures the trace having the best coverage for a particular task is marked with "-" characters around the count. When deciding which trace is the strongest for a task, we use a hidden parameter that defines how much the coverage count of a trace needs to be larger than the other traces (in this case, the value of the hidden parameter was chosen to be 4x which means that the ratio between highest count value against the next highest value in each row should be greater then 4), taking the normalization into account, for it to be considered a strong area. For example, the tuple **"0 0 0 2055 746 101"** has no strong value because 2055, which is the highest value, is not larger by the factor (less then 4x) we chose when compared to the next highest value of 746.

2.3 Single and Grouping Comparison

The traces to compare can be collected from various sources such as testing systems, loads that try to emulate customers, or real customers. Traces from different sources often tend to have very different coverage. Conversely, multiple traces originating from the same system often tend to have similar coverage. As our naming technique (for details, see 2.3.1.2   Naming Report section) looks for areas in which the trace has better coverage than all others, comparing between similar traces can result with no named area for the trace. For example, if we take the same trace twice, then the two copies, while they may be stronger than the rest of the traces in many places, will be unique in none. To alleviate the problem arising from multiple traces from the same source, we introduce a grouping comparison method.

Our tool has two multi-compare comparison methods: Single or Grouping.

2.3.1 Single Comparison
In a single comparison, all traces are compared, task by task. The raw data comparison report discussed above in 2.2   Comparing raw data from multiple data sources is common to both single and group reports. However, the Summary and Naming Report sections are different for each report type. The following sections describe their contents in the single comparison mode.

2.3.1.1   Summary Sections. Figure 4 shows the single comparison report's summary sections, which contain statistics and other high-level useful information. First, the total coverage and coverage percentage is displayed. Then, information about each trace is displayed. For each trace, the following details are reported:

- Trace number and name; for example, the first trace's name is IMSA_PMR02978.
- Normalization divisor, used for normalizing the coverage count of all tasks belonging to the trace. We need to normalize due to the different running times of each trace. The normalization divisor is automatically calculated from the coverage counts of all traces. For example, the compared tasks' count values for IMSA_PMR02978 trace in the report are the original values divided by 5.44 which means that running time of that trace was estimated to be 5.44 times larger then trace #6. Determining a normalization algorithm that will result in good naming for the traces is difficult. One can think of good reasons to have no normalization factor, or to use the time as the normalization factor, amongst other options. Having experimented with many options we chose the one that is computed by the method described in 2.4 Naming algorithm.
- Total and percentage coverage that are measured for the trace. For example, the measured coverage for IMSA_PMR02978 trace is 7 and the percentage value is 0.01.

- Strength count is the number of tasks in a particular trace that have better coverage than all the other traces. For example, the trace named IMSA_PMR02978 has one task for which its coverage is higher while IMSA_PMR02996 has six tasks with higher coverage values.
- Uniqueness count is a special case of strength count. It counts the number of tasks for which the test is the only test that covers this task. For example, the task **"0 0 0 0 0- 5 - ATREND-End_Transaction-short-0-0"** is a unique tuple. Its coverage value of the sixth trace is 5 while none of the other traces cover this task.
- #Strength-areas counts areas of tasks in a trace that have better coverage than other traces and have something in common. Names are given to those areas as be seen in the next section.

```
1
2 Summary
3 Total tasks 52560 Coverage=   65   Percentage= 0.12
4
5 Sets
6 1-IMSA_PMR02978
7   Normaliztion-Divisor=5.44 Coverage= 7  Percentage=0.01 Strength-count=1 Uniqueness-count=0 #Strength-areas=0
8 2-IMSA_PMR02996
9   Normaliztion-Divisor=1.72 Coverage=14 Percentage=0.03 Strength-count=6 Uniqueness-count=6 #Strength-areas=1
10 3-IMSA_PMR02997
11   Normaliztion-Divisor=4.32 Coverage= 7  Percentage=0.01 Strength-count=0 Uniqueness-count=0 #Strength-areas=0
12 4-IMSB_PMR02978
13   Normaliztion-Divisor=1.53 Coverage=29 Percentage=0.06 Strength-count=1 Uniqueness-count=1 #Strength-areas=0
14 5-IMSB_PMR02996
15   Normaliztion-Divisor=1.27 Coverage=30 Percentage=0.06 Strength-count=3 Uniqueness-count=2 #Strength-areas=1
16 6-IMSB_PMR02997
17   Normaliztion-Divisor=1.00 Coverage= 47 Percentage=0.09 Strength-count=24 Uniqueness-count=19 #Strength-areas= 7
```

Fig. 4. Single report comparison summary

2.3.1.2 Naming Report Section. Figure 5 shows an example of our naming method. It displays the names of the strong areas of each trace. In Figure 4 we can see that IMSA_PMR02978 has zero strength areas, IMSA_PMR02996 has one, etc. We use those strength areas for naming the traces. These names represent strong characteristics of the traces. For example, IMSB_PMR02996 has one name: "Log_Name-long-0-0 (2) (U2,S0)", while IMSA_PMR02978 has no names and IMSB_PMR02997 has seven. A name has three parts: a string which is the common substring of the strong tasks, a number between round parentheses that is the number of strong tasks in this area, and another set of round parentheses that contains the 'U' character, the number of unique tasks, the character 'S', and the number of stronger tasks in this name. A stronger task is a task with a particular trace that has better coverage than all the other traces, and at least one of its other trace's coverage is not 0.

Although not shown below it is possible to ask for a detailed report that displays the tasks that contribute to each name.

```
19  Tests by Strings strength
20
21  1-IMSA_PMR02978
22  2-IMSA_PMR02996
23      _Log_Name-long-low- (4) (U4,S0)
24  3-IMSA_PMR02997
25  4-IMSB_PMR02978
26  5-IMSB_PMR02996
27      _Log_Name-long-0-0 (2) (U2,S0)
28  6-IMSB_PMR02997
29      -medium-medium- (5) (U4,S1)
30      ATRSUSI*1 (5) (U3,S2)
31      ATREND-End_Transaction-medium- (4) (U3,S1)
32      -Express_UR_Interest-medium- (4) (U4,S0)
33      -Post_Deferred_UR_Exit-medium- (3) (U2,S1)
34      -short-low-1 (2) (U2,S0)
35      -long-high-0 (2) (U2,S0)
```

Fig. 5. Strength comparison report – naming traces

2.3.2 Group Comparison

In section 2.2 Comparing raw data from multiple data sources we described the role of the hidden parameter that defines how much the coverage count of a trace needs to be larger than the other traces. In a group comparison, the traces are allocated into groups and then the each trace is compared to its group's traces without using the above hidden parameter and to other traces by using it. The Summary and Naming Report sections are different for each report type. The following two sections describe their contents in the group comparison mode. The six traces from previous sections were grouped into two groups; the tests IMSA_PMR02978(1), IMSA_PMR02996(2), and IMSA_PMR02997(3) were allocated to the first group, and IMSB_PMR02978(4), IMSB_PMR02996(5), and IMSB_PMR02997(6) to the second.

2.3.2.1 Summary Sections. Figure 6 shows the group comparison report's summary sections, which contain statistics and other high-level useful information. First, the total coverage and coverage percentage is displayed. Then, information about each group is displayed. For each group, the following details are reported:

- Group number and name. The group's name is automatically built from the test names that are part of it. For example the first group's name is IMSA_PMR02978(1),IMSA_PMR02996(2),IMSA_PMR02997(3).
- Total and percentage coverage that were measured for the group. The group's coverage equals the union of all its trace tasks. For example, the measured coverage for the first group is 15 and the percentage value is 0.03.
- Strength count is the number of tasks in a particular group that have better coverage than all the other groups. For example, the first group has 13 tasks and the second group has 50 tasks with higher coverage values.
- Uniqueness count is a special case of strength count. It counts the number of tasks for which this is the only group that covers this task. For example,

the following task: **" - 529 - 415 254   0   0   0   ATRCMIT- Com-mit_UR-medium-0-0"** is a unique tuple, because the first three counts are for group 1 and the last three counts are for group1. The coverage value of the first group is 520 while none of the other groups cover this task.

- #Strength-areas counts areas of tasks in a group that have better coverage than other groups and have something in common. Names are given to those areas as be seen in the next section.

```
 1
 2  Summary
 3  Total tasks  52560  Coverage=    65    Percentage= 0.12
 4
 5  Groups
 6
 7  1-[ IMSA_PMR02978(1),IMSA_PMR02996(2),IMSA_PMR02997(3)]
 8        Coverage=15  Percentage=0.03  Strength-count=13  Uniqueness-count=11  #Strength-areas=2
 9
10  2-[ IMSB_PMR02978(4),IMSB_PMR02996(5),IMSB_PMR02997(6)]
11        Coverage=54  Percentage=0.10  Strength-count=50  Uniqueness-count=50 #Strength-areas=6
12
```

Fig. 6. Group report comparison summary

2.3.2.2   Naming Report Section. Figure 7 displays the names of the strong areas of each group. Figure 6 show that the first group has two strength areas, while the second group has six. We use these strength areas for naming the groups. The names represent strong characteristics of the groups. The naming method and the meaning of each part of the name are similar to the single report naming, as described in 2.3.1.2 Naming Report section.

It is possible to ask for a detailed report displaying the tasks that contribute to each name.

```
12
13  Groups by Strings strength
14
15    1-[ IMSA_PMR02978(1),IMSA_PMR02996(2),IMSA_PMR02997(3)]
16        -long-low- (6) (U6,S0)
17        ATRCMIT-Commit_UR- (5) (U5,S0)
18
19    2-[ IMSB_PMR02978(4),IMSB_PMR02996(5),IMSB_PMR02997(6)]
20        -high- (23) (U23,S0)
21        ATREND-End_Transaction- (13) (U13,S0)
22        ATRPDUE-Post_Deferred_UR_Exit- (10) (U10,S0)
23        -short-medium- (9) (U9,S0)
24        -medium-medium- (5) (U5,S0)
25        _Log_Name-long-0-0 (2) (U2,S0)
```

Fig. 7. Strength comparison report – naming of groups

2.4 Naming Algorithm

The naming algorithm is an automatic approach that gives meaningful names to each test or the customer for whom the name is special. For each test's strength area, Hole Analysis [9] finds descriptive names. The naming algorithm enables us to depict the tests' strength using only a few names, even when we have a huge amount of coverage data to process.

The naming algorithm has five major stages:

1. Builds tasks from all the traces. Each task contains a name and a tuple with normalized coverage values of each trace. For each trace there is a normalization factor which is a function of its total coverage and number of tasks that have coverage. The normalized coverage value is a result of original coverage value divided by the normalization factor. In case that running times of input traces are known, the normalization factor is computed as the ratio between lowest running time and each other running times.
2. Compares the normalized coverage values of each trace against all the other traces.
3. Creates, for each trace, a list of all the tasks for which it is strong.
4. Creates a new set of traces (one trace for each original trace) with new coverage counts, as follows:
    a. Sets the coverage values of all strong tasks to 0, which means non-covered value for the substring hole in step 5.
    b. Sets the coverage values of all the other tasks to 1, which means covered value for the substring hole in step 5.
5. Analyses substring holes [3] on the new traces. The analysis results identify areas of strength of the selected trace, compared to all other traces. It is possible to use other holes generating algorithms [4] when the model is a functional coverage model.

The result for each trace is a set of hole names that represent the areas in which this trace is better than the others. This is useful when trying to understand if this trace should be used.

3 Initial Experiences

Our first example uses sample data from *Evaluating Workloads Using Comparative Functional Coverage* [1]. A model was constructed to instrument the module calls in a data base access software component of the z/OS operating system. Upon detecting a module entry, the attributes for the model measure whether the entered module is already running on the same thread or another thread in the system, and whether any other modules are already running on the same thread or other threads.

Two test runs were compared, one from the customer and one from the test. The original analysis observed that the customer ran with significantly lower levels of concurrency than Systest. This is illustrated in Figure 8 below, taken from the earlier paper. For few tasks, a high count for no other module running on other thread (OtherOnSystem = 0) correlates to low concurrency.

Summary
Tasks: 98   # Changes: 10   Covered: 7 : 5   Coverage Percentage: 7.1%   : 5.1%

Coverage sort

Changed on bottom	Ratio: [1] / [2]

Click on a column header to sort the rows by this column

Module	OtherOnThread	OtherOnSystem	Customer	SysTest
IDAVRBF2-25-Buffer-Mgr-2	0	0	0	3
IDAVRPS1-68-PC_SS_Common_Rtn	0	0	28	0
IDAVRR48-87-call-catalog	0	0	0	1
IDAVRRM0-14-Data-Insert	0	0	0	1
IDAVRRP0-15-EndReq	0	0	62	0
IDAVRRR0-54-RRDS-processing	0	0	108	0
IDAVRRU0-42-path-upgrade	0	0	63	0
IDAVRRX0-41-path-access	0	0	242	0
IDAVRBFM-23-Buffer-Manager	0	0	191	5
IDAVRR40-06-R40-Get-path	0	0	2312	3
IDAVRAIX-86-AIX-Cleanup-Routine	0	0	0	0
IDAVRARR-67-PC-ARR	0	0	0	0

Fig. 8. Two-trace comparison showing concurrency

In contrast Figure 9 demonstrates a part of the multi-comparative functional analysis for the same two workloads. A similar conclusion regarding concurrency is identified in the highlighted line "-0-1-0-0 (25)", meaning that MeOnSystem=0, OtherOnThread=1, OtherOnSystem=0, MeNextOnThread=0 was observed 25 times. Of these values OtherOnSystem=0 is an indicator of low concurrency in the SingleModuleCustomer trace. Interestingly, the new analysis technique identified this automatically, without the need for user-generated model restrictions.

```
Tests by Strings strength

  1-SingleModuleITest_____
     -0-1-0-1-0 (4)
     IDAVRBFM-23-Buffer-Manager-0-1 (3)
     IDAVRBFM-23-Buffer-Manager-1-1 (2)
  2-SingleModuleCustomer
     -0-1-0-0 (25)
     -path-0-0- (6)
     IDAVRRU0-42-path-upgrade-0-0- (4)
     -RRDS- (4)
     IDAVRPS1-68-PC_SS_Common_Rtn-0 (3)
     IDAVRTX1-19-Term-Exit-0-0-0-1- (2)
     0-41-path-access-0-0-0-0- (2)
```

Fig. 9. Multi-trace comparison showing concurrency

We reach an additional conclusion in the "-RRDS- (4)" line. For clarity, the report information for this strength area is expanded in Figure 10. It identifies two modules that are used with more frequency by the customer workload than by the test workload.

A closer inspection of the entries for these modules reveals that they are used by the customer 15 to 30 times more frequently than the test. This significant conclusion was not previously identified when using simple comparative function coverage.

```
-RRDS- (4)
    IDAVRRQ0-55-RRDS-Put/Erase/IDALKADD-0-0-1-0-0
    IDAVRRQ0-55-RRDS-Put/Erase/IDALKADD-0-0-1-1-0
    IDAVRRR0-54-RRDS-processing-0-0-0-0-0
    IDAVRRR0-54-RRDS-processing-0-0-0-1-0
```

Fig. 10. Detail of "-RRDS-" strength

The second example compares several test runs from SysTest systems. In the traces resource manager transactions originated on the frontend system and were processed by the backend system. Traces were collected in matching sets, one from each system. Thus, both sides of any transactions were recorded in each set of matching traces. Three such sets of traces were collected for a total of six individual traces.

If each of the three measured runs was of the same workload, we would expect to see no strength areas identified because the tool would find that all the traces match for the frontend systems. Likewise, all the traces for the backend systems would match. However, the Summary section for the analysis in Figure 11 shows that several strength areas were identified. In particular, SYSA_TRACE2 and SYSB_TRACE3 indicate something unusual occurred in these runs, as evidenced by the high strength and uniqueness counts for each.

```
Summary
Total tasks  52560         Coverage=   65    Percentage= 0.12

Sets
  1-SYSA_TRACE1
        Normaliztion-Divisor=5.44     Coverage=  7      Percentage=0.01
        Strength-count=  1            Uniqueness-count=  0        #Strength-areas=  0
  2-SYSA_TRACE2
        Normaliztion-Divisor=1.72     Coverage= 14      Percentage=0.03
        Strength-count=  6            Uniqueness-count=  6        #Strength-areas=  1
  3-SYSA_TRACE3
        Normaliztion-Divisor=4.32     Coverage=  /      Percentage=0.01
        Strength-count=  0            Uniqueness-count=  0        #Strength-areas=  0
  4-SYSB_TRACE1
        Normaliztion-Divisor=1.53     Coverage= 29      Percentage=0.06
        Strength-count=  1            Uniqueness-count=  1        #Strength-areas=  0
  5-SYSB_TRACE2
        Normaliztion-Divisor=1.27     Coverage= 30      Percentage=0.06
        Strength-count=  3            Uniqueness-count=  2        #Strength-areas=  1
  6-SYSB_TRACE3
        Normaliztion-Divisor=1.00     Coverage= 47      Percentage=0.09
        Strength-count= 24            Uniqueness-count= 19        #Strength-areas=  7
```

Fig. 11. Six-trace compare summary

The Tests by Strings strength section in Figure 12 from the same report gives more detail about the strengths. Both SYSA_TRACE2 and SYSB_TRACE2 call out functions that contain "Log_Name" in the function name. An inspection of the matching detailed report (not shown) does, in fact, show these two traces are the only ones that invoke the Retrieve_Log_Name and Set_Log_Name functions. However, Figure 12 also shows a difference between the calls on these two systems. The calls on SYSA always run while other resource manager calls are in progress, but SYSB seems to run them when no other resource manager activity is occurring.

We also see that SYSB_TRACE3 contains several strength areas that indicate something very different occurred on that system. A review of the associated problem report revealed that a resource manager restart occurred on that run that did not happen on the others. This drove several functions related to that restart.

```
Tests by Strings strength

 1-SYSA_TRACE1
 2-SYSA_TRACE2
    _Log_Name-long-low- (4) (U4,S0)
 3-SYSA_TRACE3
 4-SYSB_TRACE1
 5-SYSB_TRACE2
    _Log_Name-long-0-0 (2) (U2,S0)
 6-SYSB_TRACE3
    -medium-medium- (5) (U4,S1)
    ATRSUSI*1 (5) (U3,S2)
    ATREND-End_Transaction-medium- (4) (U3,S1)
    -Express_UR_Interest-medium- (4) (U4,S0)
    -Post_Deferred_UR_Exit-medium- (3) (U2,S1)
    -short-low-1 (2) (U2,S0)
    -long-high-0 (2) (U2,S0)
```

Fig. 12. Six-trace strengths detail

Grouping the same set of traces by the originating systems and the processing systems gives us a different perspective of the data. Assuming different sets of resource manager functions are used by the originating and processing systems, we would expect to see corresponding strengths on each group. Figure 13 bears this out. The ATRCMIT-Commit_UR function is used primarily on the frontend system. Likewise, ATREND-End_Transaction and ATRPDUE-Post_Deferred_UR_Exit functions are used primarily on the backend system.

In addition to identifying functions used by one group or another, the report also makes some interesting observations. Out of the 13 strength areas in group 1, six of them contain "-long-low-". This indicates that even though there are a low number of concurrent requests being processed, the request takes a "long" time to complete. In group 2, "-high–" is observed in 23 of 50 strength areas, indicating a high number of concurrent requests.

```
Summary
Total tasks  52560          Coverage=   65    Percentage= 0.12

Groups

1-[ SYSA_TRACE1(1),SYSA_TRACE2(2),SYSA_TRACE3(3)]
            Coverage=15      Percentage=0.03   Strength-count=13
            Uniqueness-count=11        #Strength-areas=2

2-[ SYSB_TRACE1(4),SYSB_TRACE2(5),SYSB_TRACE3(6)]
            Coverage=54      Percentage=0.10   Strength-count=50
            Uniqueness-count=50        #Strength-areas=6

Groups by Strings strength

  1-[ SYSA_TRACE1(1),SYSA_TRACE2(2),SYSA_TRACE3(3)]
     -long-low- (6) (U6,S0)
     ATRCMIT-Commit_UR- (5) (U5,S0)
  2-[ SYSB_TRACE1(4),SYSB_TRACE2(5),SYSB_TRACE3(6)]
     -high- (23) (U23,S0)
     ATREND-End_Transaction- (13) (U13,S0)
     ATRPDUE-Post_Deferred_UR_Exit- (10) (U10,S0)
     -short-medium- (9) (U9,S0)
     -medium-medium- (5) (U5,S0)
     _Log_Name-long-0-0 (2) (U2,S0)
```

Fig. 13. Grouped comparison

4 Conclusions

Coverage is usually done on an aggregation of tests. The most common use of coverage is to look at what is missing from the testing so far. In previous work [1] we showed that comparative functional coverage is very useful. The work then was in the context of comparing test to customer but the same technology was later applied to other contexts such as comparing what was covered by tests to what changed in the code.

This paper extends the work to compare multiple sources. We had two separate initial motivations. The first was to understand regression buckets, a set of tests to be executed after every code change to see if bugs creep in. The problem is with the list of collected tests that need to be run when the code is modified, in that we no longer know why each test is there. We would like an automated process that looks at all the tests and tells us, in a meaningful manner, what is special about each. The second motivation was to look at workloads collected from customers and tests, and to be able to tell where the tests need to be improved. The problem is complicated by the fact that customer and test each have a number of workloads. The initial algorithm can no longer be used as we need to divide the workloads into groups to learn how each group is special. Looking at each workload by itself and aggregating the results will not work, as we explained.

An interesting side effect is that the analysis used in the paper for aggregating information can automatically find things that previously required manual intervention.

This was done by combining new code coverage techniques [2] [3] with functional coverage comparison techniques [1].

We have seen that the new reporting mechanisms are being used in ways we did not envision. For example, recently, we have seen it used when comparing coverage results between multiple regression tests. It will be interesting to see what new applications will emerge that use the multi-source comparison report.

References

1. Adler, Y., Blue, D., Conti, T., Prewitt, R., Ur, S.: Evaluating Workloads Using Comparative Functional Coverage. In: Chockler, H., Hu, A.J. (eds.) HVC 2008. LNCS, vol. 5394, pp. 84–98. Springer, Heidelberg (2009)
2. Adler, Y., Farchi, E., Klausner, M., Pelleg, D., Raz, O., Shochat, M., Ur, S., Zlotnick, A.: Automated substring hole analysis. In: ICSE 2009, pp. 203–206 (2009)
3. Adler, Y., Farchi, E., Klausner, M., Pelleg, D., Raz, O., Shochat, M., Ur, S., Zlotnick, A.: Advanced Code Coverage Analysis Using Substring Holes. In: ISSTA 2009 (2009)
4. Azatchi, H., Fournier, L., Marcus, E., Ur, S., Zohar, K.: Advanced Analysis Techniques for Cross-Product Coverage. IEEE Trans. Comput. 55(11), 1367–1379 (2006)
5. Beizer, B.: Software Testing Techniques. Van Nostrand Reinhold, New York (1990)
6. Bryce, R.C., Colbourn, C.J.: Test prioritization for pairwise interaction coverage. In: Proceedings of the 1st International Workshop on Advances in Model-Based Testing (May 2005)
7. Buchnik, E., Ur, S.: Compacting regression-suites on-the-fly. In: Proceedings of the 4th Asia Pacific Software Engineering Conference, pp. 385–394 (December 1997)
8. Grinwald, R., Harel, E., Orgad, M., Ur, S., Ziv, A.: User Defined Coverage—a Tool-Supported Methodology for Design Verification. In: Proceedings of the 35th Annual Conference on Design Automation (1998)
9. Lachish, O., Marcus, E., Ur, S., Ziv, A.: Hole Analysis for Functional Coverage Data. In: Proceedings of the 39th Conference on Design Automation (2002)
10. Marick, B.: The Craft of Software Testing: Subsystem Testing Including Object-based and Object-oriented Testing. Prentice-Hall, Englewood Cliffs (1995) ISBN 0131774115
11. Piziali, A.: Functional Verification Coverage Measurement and Analysis. Springer, Heidelberg (2004)
12. Yang, Q., Li, J.J., Weiss, D.: A survey of coverage based testing tools. In: Proceedings of the 2006 International Workshop on Automation of Software Test, AST 2006 (2006)
13. Coverage-Driven Functional Verification: Using Coverage to Speed Verification and Ensure Completeness. Verisity Design, Inc. (2001),
   `http://www.verisity.com/resources/whitepaper/`
   `coverage_driven.html` (retrieved)
14. Specman tool, `http://www.verisity.com/products/specman.html` (retrieved)
15. FoCuS tool, `http://www.alphaworks.ibm.com/tech/focus` (retrieved)

Reasoning about Finite-State Switched Systems

Dana Fisman[1,2,*] and Orna Kupferman[1]

[1] School of Computer Science and Engineering, Hebrew University, Jerusalem 91904, Israel
[2] IBM Haifa Research Lab, Haifa University Campus, Haifa 31905, Israel

Abstract. A *switched system* is composed of components. The components do not interact with one another. Rather, they all interact with the same environment, which switches one of them on at each moment in time. In standard concurrency, a component restricts the environment of the other components, thus the concurrent system has fewer behaviors than its components. On the other hand, in a switched system, a component suggests an alternative to the other components, thus the switched system has richer behaviors than its components.

We study finite-state switched systems, where each of the underlying components is a finite-state transducer. While the main challenge, namely compositionality, is similar in standard concurrent systems and in switched systems, the problems and solutions are different. In the verification front, we suggest and study an assume-guarantee paradigm for switched systems, and study formalisms in which satisfaction of a specification in all components imply its satisfaction in the switched system. In the synthesis front, we show that while compositional synthesis and design are undecidable, the problem of synthesizing a switching rule with which a given switched system satisfies an LTL specification is decidable.

1 Introduction

Concurrent systems are composed of components. Traditional concurrency theory considers two types of concurrent composition operators: *synchronous parallel composition* and *asynchronous parallel composition* (a.k.a. *interleaving*). In the former the components proceed simultaneously and in the latter their behaviors are interleaved. In both, the components not only interact with the environment but also with one another. There are, however, many natural settings in which components do not interact with one another. Rather, at each moment in time one of the components determines the behavior of the system, while the other components are ignored. Such a "switching semantics" has been well-studied in the engineering community [12,13]. In this paper, we study it for finite-state systems.

Given finite-state transducers $\mathcal{T}_1, \mathcal{T}_2, \ldots, \mathcal{T}_n$, all interacting with the same environment, we define the *switched system* $\mathcal{T}_1 \oplus \mathcal{T}_2 \oplus \cdots \oplus \mathcal{T}_n$ as a transducer that proceeds, in each moment in time, according to one of the underlying transducers.[1] There are two natural definitions for the $\oplus$ operator. In a *dormant* composition, components that are suspended are not active. That is, when a component is switched on again, it proceeds

---

* The work of this author was done as part of the Valazzi-Pikovsky Fellowship Fund.
[1] A *transducer* is an input/output finite state machine, formally defined in Section 2. We use transducers to model concurrent systems.

K. Namjoshi, A. Zeller, and A. Ziv (Eds.): HVC 2009, LNCS 6405, pp. 71–86, 2011.
© Springer-Verlag Berlin Heidelberg 2011

from the state it has reached before its suspension. In an *active* setting, components that are suspended continue their dynamics and have full observability of the environment, but their output is ignored.[2]

As an example to a dormant composition, consider a window system; at each moment in time, several windows are open and the location of the mouse determines which window is active. The other windows are inactive. We would like the window system to have the property that if a keyboard input occurs while the active window is in an "insert password" subroutine, then the char * is displayed; and if the pressed key is "enter", then the last typed chars are matched against the correct password. Note that this property should hold even if the window system switches among different window while the user types. This example shows that, even in the dormant setting, the configuration of components that are switched off should be maintained.

As an example for a switched system with an active composition, consider a network of security cameras. The cameras are located in several locations, and each camera is equipped with a software analyzing the picture. If a suspicious behavior is detected by the software, the picture is frozen until another suspicious behavior is detected. At each moment in time the output of one of the cameras is displayed at the guard's control screen. We would like to reason about the switched system and the various possible switching rules for it. For example, under an arbitrary switching rule, the system does not satisfy the property "all suspicious behaviors are detected," and it does satisfy the property "if suspicious behaviors are detected simultaneously in all locations, then at least one of them is displayed on the control screen". Also, under certain assumptions, like the configuration of the building and the location of the cameras, it is possible to synthesize a switching rule with which at least one frozen picture of a sequence of suspicious behaviors is displayed. As another example to the active composition, consider a channel TV. Obviously, broadcasting continues (but is ignored) for channels that are switched off. Using the setting of switched systems, we can reason about properties of the entire system. For example, if we are an advertising agency, we would like to synthesize an advertisement scheduling so that a viewer may not be able to avoid advertisement no matter what his switching rule is.

Finite-state switched systems, as defined above, may also serve as an abstraction of other, not necessarily finite-state, switched systems. Examples to switched systems include software systems (c.f., internet communication protocols [7,22]), mechanical systems (engines with gear transmission [8]), electrical circuits (power converters [6]), biological systems (gene regulating networks [3]), and embedded systems combining the above [2]. There has been extensive research in the control engineering community on analysis of continuous switched systems whose evolution is described by means of differential equations [12,13]. The study there focuses on properties such as stability.

---

[2] Dormant switched systems may seem similar to co-routines. A *co-routine* specifies several points in the code, referred to as *yield points*. When the scheduler is invoked, it passes control to one of the co-routines that are at their yield point. Thus, as in dormant switched systems, when a component is reinvoked it continues from the state it has reached when last invoked (rather than from the initial state as in ordinary routines). Unlike switched systems, however, the components do have control on when the scheduler is invoked. Anyway, the theoretical aspects of co-routers have not been investigated.

The theory of verification considers other type of properties, those expressible in temporal logic. Thus, considering abstraction of continuous systems enables reasoning about other aspects of systems. For example, consider a cell phone that may move among different receiving zones. This is a popular example for continuous switched systems [14], yet many properties of the system can be specified in temporal logic. For example, we would like to check that whenever a *network available* signal appears, it stays valid as long as the cell phone does not change its location, and that if a call was issued, then eventually either the network is no longer available or the call gets to the target phone. These properties should hold even if the cell phone changes its location. Such a setting corresponds to the dormant composition – the operation of the cell phone in a particular zone is a component (note that the cell phone operates differently in different zones), and transiting among the zones correspond to switching.

The above examples highlight the *Gestalt principle*, which is accepted in the study of continuous switched systems. According to this principle "the sum of the whole is greater than its parts". For example, a continuous switched system may be stable even though its underlying components are not stable, and vice versa, a continuous switched system may be unstable even though its underlying components are stable [13]. This is in contrast with standard concurrency, where the concurrent system has fewer behaviors than its components. The fact that the composed system has fewer behavior than its components has played a central role in compositional reasoning. As shown in [1], both synchronous and asynchronous parallel compositions can be seen as intersection of the enhanced language of its components. Further classes of parallel compositions have been studied in [1]. They all, however, convey a notion of intersection between languages. As we shall show, our dormant and active compositions convey a notion of union rather than intersection. Thus, general ideas and patterns that are applicable in the study of standard concurrency cannot be applied in the setting of switched systems.[3]

As in standard concurrency, composing finite-state transducers via active or dormant compositions involves an exponential blowup. Thus, the main challenge in reasoning about finite-state switched systems is *compositional reasoning*, i.e., reducing reasoning about a concurrent system to reasoning about its individual components. While the main challenge, namely compositionality, is similar in switched systems and standard concurrent systems, the problems and solutions are different. We start by studying the compositional model-checking problem for switched systems. As noted above, an algorithm that constructs the switched system explicitly is possible. We show that the space complexity of LTL model checking is polynomial in the size of the underlying components, thus the exponential blow-up that the construction of an explicit switched system involves cannot in general be avoided.

---

[3] By letting the variables of the different components be disjoint, it is possible to model the dormant and active compositions using known synchronous and asynchronous composition operators. Such a modeling, however, is less clean, and hides the switching mechanism. In [15], Mayer and Stockmeyer studied regular expressions extended with a *shuffle* operator on words, which interleaves its operands. As we show later, the shuffle operator corresponds to a dormant composition between closed systems. Our setting here is richer, as it considers open systems. We also study different problems than those studied in [15].

Note that since a component may be switched on forever, a required condition for a switched system to satisfy a property is that all the underlying components satisfy it. For some properties, this is also a sufficient condition, giving rise to a simple compositional model-checking procedure for them that avoids this blow-up. We characterize such properties for the dormant composition by means of *regular counting* properties, and conclude that, unfortunately, most interesting properties cannot enjoy this simple procedure. We then describe an assume-guarantee paradigm for switched systems [16], which enables us to reason about a switched system (with respect to all LTL specifications) by reasoning about its components, and often avoid this blow up. Formally, a transducer $\mathcal{T}$ satisfies the assume-guarantee specification $\langle \varphi, \psi \rangle$, for LTL specifications φ and ψ, and a composition operator $\oplus$, if for all transducers $\mathcal{T}'$, if $\mathcal{T} \oplus \mathcal{T}'$ satisfies φ, then $\mathcal{T} \oplus \mathcal{T}'$ also satisfies ψ. We study the problem of checking assume-guarantee specifications and show that it is PSPACE-complete. Unlike traditional concurrency, the problem cannot be reduced to checking whether $\mathcal{T}$ satisfies $\varphi \rightarrow \psi$ [16]. Indeed, the latter reduction depends on the fact that compositions that have $\mathcal{T}$ as a component have fewer behaviors than $\mathcal{T}$, which does not hold for switched systems. We show that for switched systems checking whether $\mathcal{T}$ satisfies the assume-guarantee specification $\langle \varphi, \psi \rangle$ has the flavor of checking validity of $\varphi \rightarrow \psi$. This is due to monotonicity that does hold for switched systems as well.

The model-checking problem checks whether a given switched system satisfies a specification under arbitrary switching. A more ambitious goal is synthesis – the automatic construction of systems from specifications. In the switched setting, given LTL specifications $\varphi_1, \varphi_2, \ldots, \varphi_n$, and ψ, and a composition operator $\oplus$, the *compositional-realizability* problem is to decide whether there are transducers $\mathcal{T}_1, \mathcal{T}_2, \ldots, \mathcal{T}_n$ such that $\mathcal{T}_i$ satisfies φ_i for all $1 \leq i \leq n$, and $\mathcal{T}_1 \oplus \mathcal{T}_2 \oplus \cdots \oplus \mathcal{T}_n$ satisfies ψ. On the negative side, we show that, as with standard concurrency [18] compositional-realizability is undecidable. Sometimes, the details of the switching mechanism are known and may be controlled. On the positive side, we study the problem of synthesizing a *switching rule* according to which the switching system satisfies a specification. We show that the problem has the same flavor as the standard LTL control problem, and is 2EXPTIME-complete [17]. The solution to the problem, however, is different, as the synthesized switched rule does not disable transitions, as is the case in usual control. Rather, it chooses the component that is switched on.

2   Transducers and Switched Systems

Let I and O be finite sets of input and output signals. Let Σ_I and Σ_O denote the sets 2^I and 2^O, respectively. Let Σ_{IO} denote the set $\Sigma_I \times \Sigma_O$. A *transducer* is an automaton on finite words over the alphabet Σ_I in which each state is associated with a letter in the alphabet Σ_O. A transducer does not have an acceptance condition. The intuition is that the transducer models an *open system* that interacts with its environment. In each moment in time the system reads a set $i \in \Sigma_I$ of input signals that are valid in this moment, changes its state according to i, and outputs a set $o \in \Sigma_O$ of output signals that are valid in the new state.

Formally, a transducer is a tuple $T = \langle \Sigma_I, \Sigma_O, S, \theta, \eta, L \rangle$, where S is a set of states, $\theta : \Sigma_I \to S$ is an initialization function mapping the first input letter to an initial state, $\eta : S \times \Sigma_I \to S$ is a transition function, and $L : S \to \Sigma_O$ is a labeling function. The *run* of T on an input sequence $i_0 \cdot i_1 \cdot i_2 \cdots \in \Sigma_I^\omega$ is the sequence $s_0, s_1, s_2, \ldots$ of states for which $s_0 = \theta(i_0)$ and $s_{j+1} = \eta(s_j, i_{j+1})$ for all $j \geq 0$. A computation $w \in \Sigma_{IO}^\omega$ is *generated* by T if $w = (i_0, o_0) \cdot (i_1, o_1) \cdot (i_2, o_2) \cdots$ is such that the run $s_0, s_1, s_2, \ldots$ of T on $i_0 \cdot i_1 \cdot i_2 \cdots$ satisfies $o_j = L(s_j)$ for all $j \geq 0$. We refer to the set of computations generated by T as the *language* of T and denote it $\mathcal{L}(T)$. Note that T is responsive and deterministic (that is, it suggests exactly one successor state for each input letter), and thus T has a single run, generating a single computation, on each input sequence.

A *switched system* is composed of several components. Each component is an open system that interacts with the environment. The components do not interact with each other. Rather, they all interact with the environment, but only one component, chosen by the environment, is *switched on* at a given moment. The other components are *suspended*. We define two types of compositions between transducers. In a *dormant* composition, components that are suspended are not active. That is, when a component is switched on again, it proceeds from the state it has reached in the last time it was switched on. In an *active* setting, components that are suspended continue their dynamics and have full observability of the environment, but their output is ignored.

We formalize the two types of compositions below. For simplicity we assume systems with two components. The generalization to any finite number of components is straightforward. Let $T_1 = \langle \Sigma_I, \Sigma_O, S_1, \theta_1, \eta_1, L_1 \rangle$ and $T_2 = \langle \Sigma_I, \Sigma_O, S_2, \theta_2, \eta_2, L_2 \rangle$ be two transducers. We define the *dormant* and *active* switched systems with components T_1 and T_2, denoted $T_1 \oplus T_2$, and $T_1 \oplus T_2$, respectively, as the transducer $\langle \Sigma_{I'}, \Sigma_O, S, \theta, \eta, L \rangle$, defined as follows.

- $\Sigma_{I'} = \Sigma_I \times \{1, 2\}$. The $\{1, 2\}$ component of an input letter indicates which component will be switched on in the next cycle. We use $\langle i, who \rangle$ to refer to a letter in $\Sigma_{I'}$ where $i \in \Sigma_I$ and $who \in \{1, 2\}$. We can think of who as a fresh input signal defined over the domain $\{1, 2\}$.

- $S = S_1 \times S_2 \times \{1, 2\}$. That is, a state in the switched system is composed of the states of T_1 and T_2, and a flag indicating the component that is currently switched on. This component generates the current output.

  In the dormant composition, it is technically convenient to add to S_1 and S_2 a special state s_{init}, for components that have never been activated.

- The initialization function θ is defined as follows.
  - ⓓ In the dormant composition, the component that has never been switched on waits in the special state s_{init} until it is switched on for the first time. Accordingly, $\theta(\langle i, 1 \rangle) = \langle \theta_1(i), s_{init}, 1 \rangle$ and $\theta(\langle i, 2 \rangle) = \langle s_{init}, \theta_2(i), 2 \rangle$.
  - ⓐ In an active composition, the component that is not switched on proceeds as if it was active. Thus, $\theta(\langle i, who \rangle) = \langle \theta_1(i), \theta_2(i), who \rangle$.

- The transition function η is defined according to the type of composition as follows. Consider a state $\langle s_1, s_2, k \rangle \in S$ and an input letter $\langle i, who \rangle \in \Sigma_{I'}$.
  - ⓓ In a dormant composition, the component that is suspended stays in its current state until it is switched on again. Thus,

$$\eta(\langle s_1, s_2, k \rangle, \langle i, who \rangle) = \begin{bmatrix} \langle \eta_1(s_1, i), s_2, who \rangle & \text{if } who = 1 \\ \langle s_1, \eta_2(s_2, i), who \rangle & \text{if } who = 2 \end{bmatrix}$$

In addition, for $who \in \{1, 2\}$, we have $\eta_{who}(s_{init}, i) = \theta_{who}(i)$.

- In an active composition, the component that is suspended proceeds as if it was active. Thus,

$$\eta(\langle s_1, s_2, k \rangle, \langle i, who \rangle) = \langle \eta_1(s_1, i), \eta_2(s_2, i), who \rangle.$$

– For all states $\langle s_1, s_2, k \rangle \in S$, we have $L(\langle s_1, s_2, k \rangle) = L_k(s_k)$. That is, the output of the current state is determined by the component that is switched on.

Note that the underlying transducers T_1 and T_2 do not have who in their set of input signals. Thus, a component does not know whether it is switched on or not, and its behavior does not depend on this information.

A specification to the switched-system is over the set $I \cup O$ of signals. By allowing specifications to refer also to the signal who, we can easily restrict attention to compositions in which assumptions on the switching can be made. Formally, since our specification formalism is linear, we can replace a specification ψ over $I \cup O$ by the specification $\psi_{fair} \rightarrow \psi$, where ψ_{fair} is a formula over who describing assumptions on the switching. We will elaborate on the extended setting for problems studied in the following sections.

2.1   The Input-Output Language of a Switched System

Recall that the language $\mathcal{L}(T)$ of a transducer T is defined over the alphabet Σ_{IO}. Accordingly, $\mathcal{L}(T_1 \oplus T_2)$ refers also to the input signal who, which we often want to abstract. For a switched system, we also define the *IO-language* of $T_1 \oplus T_2$, denoted $\mathcal{L}_{IO}(T_1 \oplus T_2)$, which is obtained by projecting $\mathcal{L}(T_1 \oplus T_2)$ on Σ_{IO} (that is, ignoring the $\{1, 2\}$ component).

In Lemma 1 below we show that natural properties of the interleaving operator used in standard concurrent composition apply also to switched systems. On the other hand, it is not hard to see that unlike the case of interleaving, it is not necessarily the case that $\mathcal{L}_{IO}(T_1 \oplus T_2) \subseteq \mathcal{L}(T_1)$ or $\mathcal{L}_{IO}(T_1 \oplus T_2) \subseteq \mathcal{L}(T_2)$.

Lemma 1. *Let $\oplus \in \{\mathbb{O}, \bullet\}$ be a composition operator. For all transducers T_1, T_2, and T_3, the following hold.*

- *Commutativity: $\mathcal{L}(T_1 \oplus T_2) = \mathcal{L}(T_2 \oplus T_1)$.*
- *Associativity: $\mathcal{L}_{IO}((T_1 \oplus T_2) \oplus T_3) = \mathcal{L}_{IO}(T_1 \oplus (T_2 \oplus T_3))$.*
- *Monotonicity: If $\mathcal{L}(T_1) \subseteq \mathcal{L}(T_2)$ then $\mathcal{L}_{IO}(T_1 \oplus T_3) \subseteq \mathcal{L}_{IO}(T_2 \oplus T_3)$ for all T_3.*

It is not hard to see that, when restricted to their IO-languages, the dormant and active compositions corresponds to the *shuffle* and *merge* of languages. For two words $u, v \in \Sigma^\omega$, let

- $u \mathbb{O} v = \{u_1 v_1 u_2 v_2 u_3 v_3 \cdots \mid u_i, v_i \in \Sigma^*, u = u_1 u_2 u_3 \cdots \text{ and } v = v_1 v_2 v_3 \cdots\}$
- $u \bullet v = \{u_1 v_2 u_3 v_4 \cdots \mid u_i, v_i \in \Sigma^*, |u_i| = |v_i|, u = u_1 u_2 u_3 \cdots \text{ and } v = v_1 v_2 v_3 \cdots\}$

Thus, $u \oplus v$ shuffles the letters of u and v by interleaving subwords of u and v, whereas $u \bullet v$ merges u and v by locating in each position i the i-th letter of either u or v. Note that since the subwords v_i and u_i may be empty, we have that u and v are members of $u \oplus v$, and similarly for $u \bullet v$. The standard concurrency operator, a.k.a *interleaving*, is often confused with shuffle, though its operation is different. Indeed, as shown in [1], since interleaving is applied to components that own variables, it corresponds to conjunction of the enhanced language of its components [1]. This is not valid for the shuffle operation. The shuffle and merge operators naturally extends to languages. In Figure 1, we demonstrate the application of the shuffle and merge operators on some languages.

L_1	L_2	$L_1 \oplus L_2$	$L_1 \bullet L_2$
0^ω	1^ω	$(0+1)^\omega$	$(0+1)^\omega$
$0^\omega + 1^\omega$	$0^\omega + 1^\omega$	$(0+1)^\omega$	$(0+1)^\omega$
$(01)^\omega$	$(10)^\omega$	$((01)+(10))^\omega$	$(0+1)^\omega$
$(01)^\omega$	$(01)^\omega$	$0((01)+(10))^\omega$	$(01)^\omega$
0^ω	0^*10^ω	$0^\omega + 0^*10^\omega$	$0^\omega + 0^*10^\omega$
0^*10^ω	0^*10^ω	$0^*10^\omega + 0^*10^*10^\omega$	$0^\omega + 0^*10^\omega + 0^*10^*10^\omega$
$0^*1(0+1)^\omega$	$0^*1(0+1)^\omega$	$0^*1(0+1)^\omega$	$(0+1)^\omega$
$0^+1(0+1)^\omega$	$0^+1(0+1)^\omega$	$0^+1(0+1)^\omega$	$0(0+1)^\omega$
$(0+1)^*0^\omega$	$(0+1)^*0^\omega$	$(0+1)^*0^\omega$	$(0+1)^*0^\omega$
$(0^*1)^\omega$	$(0^*1)^\omega$	$(0^*1)^\omega$	$(0+1)^\omega$

Fig. 1. Shuffle and merge of languages

The definition of the dormant and active composition immediately implies their correspondence to the shuffle and merge operators. Formally, we have the following.

Lemma 2. Let T_1 and T_2 be two transducers. Then, $\mathcal{L}_{IO}(T_1 \oplus T_2) = \mathcal{L}_{IO}(T_1) \oplus \mathcal{L}_{IO}(T_2)$ and $\mathcal{L}_{IO}(T_1 \bullet T_2) = \mathcal{L}_{IO}(T_1) \bullet \mathcal{L}_{IO}(T_2)$.

In [15] it was shown that the shuffle operator provides succinctness in the sense that there exist languages that can be described exponentially more succinctly by using shuffle.[4] We show that the results extends for dormant composition and holds for active composition as well.

Theorem 1. Let $n \in \mathbb{N}$. There are transducers $T_1, \ldots, T_n$ such that the size of T_i is $O(1)$, and there is no transducer T with less than 2^{n-1} states such that $\mathcal{L}(T) = \mathcal{L}_{IO}(T_1 \oplus T_2 \oplus \cdots \oplus T_n)$.

The idea of the proof is to show that for any $n \in \mathbb{N}$ the set of all words over the alphabet $\Gamma_n = \{1, \ldots, n, \#\}$ in which each letter from $\{1, 2, \ldots, n\}$ appears at most once can be expressed as an active or dormant compositions of n transducers. By [15], this language cannot be generated by a transducer with less than 2^{n-1} states. The full proof is given in the full version of the paper.

---

[4] [15] refers to shuffle also as interleaving. Their definition, however, corresponds to shuffle as defined above.

3   Compositional Model Checking

The model-checking problem for a switched system is to decide, given transducers $\mathcal{T}_1, \ldots, \mathcal{T}_n$, a composition operator $\oplus \in \{\oplus, \bullet\}$, and an LTL formula ψ, whether the switched system $\mathcal{T}_1 \oplus \cdots \oplus \mathcal{T}_n$ satisfies ψ. Note that the formulation of the problem has an implicit universal quantification and the switched system has to satisfy the specification under arbitrary switching. As with the interleaving operator, it is possible to construct $\mathcal{T}_1 \oplus \cdots \oplus \mathcal{T}_n$ and model check it. As shown in Theorem 1, however, such a construction may involve an exponential blow up. Assume-guarantee reasoning avoids the blowup by inferring satisfaction of specifications in the composed system from satisfaction of specifications in the underlying components [16].

Note that since a component may be switched on forever, a required condition for a switched system to satisfy a property is that all the underlying components satisfy it. For some properties, this is also a sufficient condition, giving rise to a simple compositional model-checking procedure for them. In Section 5, we characterize such properties for the active composition. Since most interesting properties do not satisfy the characterization, we describe, in this section, an assume-guarantee paradigm for switched systems for arbitrary properties. We first show that, as with interleaving, the blow-up that the construction of $\mathcal{T}_1 \oplus \cdots \oplus \mathcal{T}_n$ involves cannot be avoided. We do so by analyzing the system-complexity of the model-checking problem, namely the complexity of the system in terms of the size of the underlying components, assuming the specification is fixed.

Theorem 2. *The system complexity of the LTL model-checking problem of switched systems is PSPACE-complete.*

Remark 1. The key to the PSPACE-hardness result is the fact that even though the components interact with the environment one at a time, they resume their interaction from a state that has to be maintained (either the state they have reached in the last time they were switched on, in a dormant composition, or the state they have reached in their silent interaction, in an active composition). A substantially different type of composition is one in which interaction is resumed from a fixed state. Then, it is possible to define the state space of the switched systems as a union of the underlying state spaces, and the system complexity of the LTL model-checking problem is NLOGSPACE complete. Fixing a state from which dynamics is resumed is even more crucial in the infinite-state setting. For example, reachability in o-minimal hybrid systems is decidable only when each discrete control state has a single initial value for the continues elements [11]. Obviously, however, resuming the interaction from a fixed state is a much weaker composition mechanism.

Remark 2. In [15], Mayer and Stockmeyer studied the complexity of membership and inequality for regular expressions extended with the shuffle operator, which as we discussed previously provides the dormant composition operator in the setting of closed system. They showed that membership is NP-complete and inequality is EXPSPACE-complete. Since equivalence is two-sided inclusion and since model checking amounts to inclusion (the language of the system should be contained in the language of the formula), their results imply that model checking of closed system restricted to finite words can be done in EXPSPACE. As Theorem 2 shows, the special case of the $A \subseteq B$ problem in which only B uses shuffle is easier, and is in PSPACE, even for the case of open systems and infinite words. Indeed, the lower bound proof in [15] uses shuffle in both sides.

We are now ready to describe an assume-guarantee paradigm for switched systems.

Definition 1. *Let $\mathcal{T}$ be a transducer. Let φ_1 and φ_2 be temporal logic formulas. Let $\oplus \in \{ \mathbb{O}, \bullet \}$ be a composition operator. We say that $\langle \varphi_1 \rangle \mathcal{T} \oplus \langle \varphi_2 \rangle$ if for every $\mathcal{T}'$, we have that $\mathcal{T} \oplus \mathcal{T}' \models \varphi_1$ implies $\mathcal{T} \oplus \mathcal{T}' \models \varphi_2$. When $\oplus$ is clear from the context, we simply write $\langle \varphi_1 \rangle \mathcal{T} \langle \varphi_2 \rangle$.*

Let $\mathcal{T}_1$ and $\mathcal{T}_2$ be two transducers, and let φ_1, φ_2, and φ_3 be LTL formulas. Below are two typical assume-guarantee rules, for a composition operator $\oplus \in \{ \mathbb{O}, \bullet \}$ (as with the known composition semantics, many more rules exist [16]).

$$\frac{\langle \varphi_1 \rangle \mathcal{T}_1 \langle \varphi_2 \rangle \quad \langle \varphi_2 \rangle \mathcal{T}_2 \langle \varphi_3 \rangle}{\langle \varphi_1 \rangle \mathcal{T}_1 \oplus \mathcal{T}_2 \langle \varphi_3 \rangle} \qquad\qquad \frac{\langle true \rangle \mathcal{T}_1 \langle \varphi_1 \rangle \quad \langle true \rangle \mathcal{T}_2 \langle \varphi_2 \rangle}{\langle true \rangle \mathcal{T}_1 \oplus \mathcal{T}_2 \langle \varphi_1 \wedge \varphi_2 \rangle}$$

Consider for example the left rule. To see that this rule is sound, note that, by definition, for every $\mathcal{T}'$ we have (1) if $\mathcal{T}_1 \oplus \mathcal{T}' \models \varphi_1$ then $\mathcal{T}_1 \oplus \mathcal{T}' \models \varphi_2$ and (2) if $\mathcal{T}_2 \oplus \mathcal{T}' \models \varphi_2$ then $\mathcal{T}_2 \oplus \mathcal{T}' \models \varphi_3$. In particular, for every $\mathcal{T}''$ we have that (1) holds when $\mathcal{T}'$ is $\mathcal{T}_2 \oplus \mathcal{T}''$ and (2) holds when $\mathcal{T}'$ is $\mathcal{T}_1 \oplus \mathcal{T}''$. Hence, for every $\mathcal{T}''$, if $\mathcal{T}_1 \oplus \mathcal{T}_2 \oplus \mathcal{T}'' \models \varphi_1$ then $\mathcal{T}_1 \oplus \mathcal{T}_2 \oplus \mathcal{T}'' \models \varphi_2$ and if $\mathcal{T}_1 \oplus \mathcal{T}_2 \oplus \mathcal{T}'' \models \varphi_2$ then $\mathcal{T}_1 \oplus \mathcal{T}_2 \oplus \mathcal{T}'' \models \varphi_3$. Hence, $\langle \varphi_1 \rangle \mathcal{T}_1 \oplus \mathcal{T}_2 \langle \varphi_3 \rangle$. Thus, the rule is sound. Similar reasoning applies for the right rule.

For the standard concurrent composition operator, interleaving, we have that $\langle \varphi_1 \rangle \mathcal{T} \langle \varphi_2 \rangle$ iff $\mathcal{T} \models \varphi_1 \rightarrow \varphi_2$. Thus, it is possible to reduce checking of an assume-guarantee specification to LTL model checking. This simple reduction relies on the fact that the language of a concurrent system is contained in the languages of its underlying components. This fact is not valid for switched systems. Instead, we should check the $\varphi_1 \rightarrow \varphi_2$ implication in a richer context:

Lemma 3. *Let φ and ψ be LTL formulas, $\oplus \in \{ \mathbb{O}, \bullet \}$, and $\mathcal{T}$ a transducer. Then, $\langle \varphi_1 \rangle \mathcal{T} \oplus \langle \varphi_2 \rangle$ iff for every transducer $\mathcal{T}'$, we have $\mathcal{T} \oplus \mathcal{T}' \vdash \varphi_1 \rightarrow \varphi_2$.*

In Lemma 1, we have shown that the operators $\mathbb{O}$ and $\bullet$ are monotone. Thus, checking $\mathcal{T} \oplus \mathcal{T}' \models \varphi_1 \rightarrow \varphi_2$ for every $\mathcal{T}'$, can be reduced to checking $\varphi_1 \rightarrow \varphi_2$ in the composition of $\mathcal{T}$ with the most challenging $\mathcal{T}'$, namely one whose language is $\Sigma_{\mathrm{IO}}{}^\omega$. Note that the monotonicity property also implies that if $\mathcal{L}(\mathcal{T}_1') = \mathcal{L}(\mathcal{T}_2')$, then $\mathcal{L}_{\mathrm{IO}}(\mathcal{T}_1' \oplus \mathcal{T}) = \mathcal{L}_{\mathrm{IO}}(\mathcal{T}_2' \oplus \mathcal{T})$. Thus, any transducer whose language is $\Sigma_{\mathrm{IO}}{}^\omega$ will do. Since a deterministic transducer generates a single computation for each input sequence, a transducer whose language is $\Sigma_{\mathrm{IO}}{}^\omega$ has to be nondeterministic. Let $\mathcal{U}$ be the nondeterministic transducer that has $|\Sigma_{\mathrm{O}}|$ states, all of them are initial, and for which each state has transitions, on all input letter in Σ_{I}, to all other states. It is easy to see that $\mathcal{L}(\mathcal{U}) = \Sigma_{\mathrm{IO}}{}^\omega$, and that the definitions of the composition operators in Section 2 extends to a composition with a nondeterministic transducer in a straightforward way.

Lemma 4. *Let φ be an LTL formula, $\oplus \in \{\mathbb{O}, \bullet\}$, $\mathcal{T}$ be a transducer, and $\mathcal{U}$ a transducer such that $\mathcal{L}(\mathcal{U}) = \Sigma_{IO}{}^\omega$. Then $\mathcal{T} \oplus \mathcal{U} \models \varphi$ iff for every $\mathcal{T}'$ we have $\mathcal{T} \oplus \mathcal{T}' \models \varphi$.*

Corollary 1. *Let φ, ψ be LTL formulas, $\oplus \in \{\mathbb{O}, \bullet\}$, $\mathcal{T}$ be a transducer, and $\mathcal{U}$ a transducer such that $\mathcal{L}(\mathcal{U}) = \Sigma_{IO}{}^\omega$. Then $\langle \varphi \rangle \mathcal{T} \oplus \langle \psi \rangle$ iff $\mathcal{T} \oplus \mathcal{U} \models \varphi \rightarrow \psi$.*

Theorem 3. *Model checking assume-guarantee specifications of switched systems is PSPACE-complete.*

Proof. As discussed above, for every transducer $\mathcal{T}$, LTL formulas φ_1 and φ_2, and a composition operator $\oplus$, we have that $\langle \varphi_1 \rangle \mathcal{T} \langle \varphi_2 \rangle$ iff $\mathcal{T} \oplus \mathcal{U} \models \varphi_1 \rightarrow \varphi_2$. Membership in PSPACE then follows from the fact that checking the latter requires space that is polynomial in φ_1 and φ_2 and logarithmic in $|\mathcal{T}| \cdot |\Sigma_O|$. The lower bound follows from the PSPACE hardness of the validity problem for LTL. Indeed, φ is valid iff $\langle true \rangle \mathcal{U} \langle \varphi \rangle$. Note that validity of LTL is PSPACE-hard already for a fixed number of propositions, thus we can consider $\mathcal{U}$ to be of a fixed size, and by classifying all the propositions as input signals, $\mathcal{U}$ is also deterministic. Thus, PSPACE-hardness holds already for deterministic transducers. $\square$

For an arbitrary switching rule, the IO-language of the composition $\mathcal{T} \oplus \mathcal{U}$ is Σ_{IO}, thus $\mathcal{T} \oplus \mathcal{U} \models \varphi_1 \rightarrow \varphi_2$ iff the implication $\varphi_1 \rightarrow \varphi_2$ is valid. Things become more interesting when assumptions on the switching are made. If, for example, $\mathcal{T} \models \mathsf{GF}\,grant_1 \rightarrow \mathsf{GF}\,grant_2$, then $\langle \mathsf{GF}\,grant_1 \rangle \mathcal{T} \langle \mathsf{GF}\,grant2 \rangle$ in a fair switching in which all components are switched on infinitely often even though $\mathsf{GF}\,grant_1 \rightarrow \mathsf{GF}\,grant_2$ may not be valid. As discussed in Section 2, such assumptions are easy to make by augmenting the specification by a precondition over *who*.

4   Synthesis of a Switching Rule

In this section we show how to synthesize a switching rule with which the composition of a given transducers satisfies a desired LTL property. Before we do so, we show that the harder problems of compositional realizability and compositional design are undecidable.

4.1   Undecidable Problems

Given LTL specifications $\varphi_1, \varphi_2, \ldots, \varphi_n$, and ψ, and a composition operator $\oplus$, the *compositional-realizability* problem is to decide whether there are transducers $\mathcal{T}_1, \mathcal{T}_2, \ldots, \mathcal{T}_n$ such that $\mathcal{T}_i$ satisfies φ_i for all $1 \leq i \leq n$, and $\mathcal{T}_1 \oplus \mathcal{T}_2 \oplus \cdots \oplus \mathcal{T}_n$ satisfies ψ. In [18] it was shown that compositional realizability is undecidable where $\oplus$ is the synchronous parallel composition. It was further shown that if, however, the processes admit a piplelined architecture the problem is decidable. In this section we show that for switched systems, though the architecture is extremely simple, compositional realizability is undecidable for both dormant and active compositions.

The *compositional-design* problem is to decide whether every switched system $\mathcal{T}_1 \oplus \mathcal{T}_2 \oplus \cdots \oplus \mathcal{T}_n$ such that $\mathcal{T}_i$ satisfies φ_i for all $1 \leq i \leq n$, also satisfies ψ. The problems of compositional-realizability and compositional design are strongly connected. Indeed, in a setting in which the formulas φ_i are realizable, the answer to the compositional-realizability problem with input $\varphi_1, \ldots, \varphi_n, \psi$ is 'yes' iff there exist transducers $\mathcal{T}_1, \mathcal{T}_2$, $\ldots, \mathcal{T}_n$ such that $\mathcal{T}_i$ satisfies φ_i for all $1 \leq i \leq n$, and $\mathcal{T}_1 \oplus \mathcal{T}_2 \oplus \cdots \oplus \mathcal{T}_n$ satisfies ψ. The latter holds iff the answer to the compositional-design problem with input $\varphi_1, \ldots, \varphi_n, \neg\psi$ is 'no'.

Theorem 4. *The compositional realizability and design problems are undecidable.*

By the above, it suffices to show that compositional-realizability problem is undecidable. The problem of compositional-realizability for standard concurrency was shown to be undecidable by Pnueli and Rosner in [18]. The key to their undecidability proof is an architecture of two processes that do *not* communicate with one another. Such lack of communication exists also in our setting, and enables an adoption of their proof with some minor adjustments.

4.2 Synthesis of a Switching Rule

Recall that the model-checking problem checks whether a switched system satisfies a specification under arbitrary switching or a switching that satisfies some assumption. Sometimes the details of the switching mechanism are known and may be controlled. In this section we study the problem of deciding, given transducers $\mathcal{T}_1, \ldots, \mathcal{T}_n$ and a specification φ, whether there is a *switching rule* according to which the switched system $\mathcal{T}_1 \oplus \cdots \oplus \mathcal{T}_n$ satisfies φ, and the problem of synthesizing such a rule in case the answer is positive.[5]

We model a switching rule by a transducer S with input alphabet Σ_I and output alphabet $\{1, 2\}$. Consider transducers $\mathcal{T}_1, \mathcal{T}_2, \ldots, \mathcal{T}_n$, a composition operator $\oplus \in \{\mathbf{\Phi}, \mathbb{O}\}$, and a switching rule S. The switched system $\mathcal{T}_1 \oplus \mathcal{T}_2 \oplus \cdots \oplus \mathcal{T}_n$ with switching rule S has input in Σ_I (rather than in $\Sigma_I \times \{1, 2\}$) and the component that is switched on after reading an input sequence $w \in \Sigma_I^*$ is determined by the output of the state of S after reading w. Formally, let $\mathcal{T}_1 = \langle \Sigma_I, \Sigma_O, S_1, \theta_1, \eta_1, L_1 \rangle$, $\mathcal{T}_2 = \langle \Sigma_I, \Sigma_O, S_2, \theta_2, \eta_2, L_2 \rangle$, and $S = \langle \Sigma_I, \{1, 2\}, S, \theta, \eta, L \rangle$. Then, the switched system with components $\mathcal{T}_1$ and $\mathcal{T}_2$, and switching rule S, is the transducer $\langle \Sigma_I, \Sigma_O, S', \theta', \eta', L' \rangle$, defined as follows:

- $S' = S_1 \times S_2 \times \{1, 2\} \times S$. Intuitively, the switched system is identical to the one without the switching rule, only that the *who* element is determined by the switching rule rather than by the environment.

- $\theta'(i) = \langle \theta_1(i), \theta_2(i), L(\theta(i)), \theta(i) \rangle$. That is, the initialization function maps each state component according to the respective initialization function, and determines the next state to be the output of the switching rule on the first input.

- The transition function η is defined according to the type of composition as follows. Consider a state $\langle s_1, s_2, k, s \rangle \in S'$ and a letter $i \in \Sigma_I$.

---

[5] A recent work [24] advocates the use of ω-regular languages over the alphabet of subcomponents identifiers for describing switching constraints even for continuous switched systems.

$$\text{①}\ \eta(\langle s_1, s_2, k, s\rangle, i) = \begin{bmatrix} \langle \eta_1(s_1, i), s_2, L(s), \eta(s, i)\rangle & \text{if } L(s) = 1 \\ \langle s_1, \eta_2(s_2, i), L(s), \eta(s, i)\rangle & \text{if } L(s) = 2. \end{bmatrix}$$

$\text{●}\ \eta(\langle s_1, s_2, k, s\rangle, i) = \langle \eta_1(s_1, i), \eta_2(s_2, i), L(s), \eta(s, i)\rangle.$

– For all $\langle s_1, s_2, k, s\rangle \in S'$, we have $L(\langle s_1, s_2, k, s\rangle) = L_k(s_k)$.

The solution to the switching-rule synthesis problem involves automata on infinite trees (see [17] or the full version of the paper).

When we synthesize a switching rule, we are given the transducers $\mathcal{T}_1$ and $\mathcal{T}_2$, and the transducer we are after only has to generate an infinite sequence over $\{1, 2\}$. The setting is then similar to the *control* problem for LTL [17]. Unlike the solution there, however, here the controller does not disable transitions. Rather, it determines which component should be active at each moment in time.

Theorem 5. *The switching-rule synthesis problem for LTL is 2EXPTIME-complete.*

Proof. Consider an LTL formula ψ. Let $\mathcal{A}_\psi = \langle \Sigma_O, Q, q_0, \delta, \alpha \rangle$ be a deterministic parity word automaton (DPW) recognizing ψ. We define a deterministic parity tree automaton (DPT) $\mathcal{A}_{\forall\psi}^{\mathcal{T}_1, \mathcal{T}_2}$ that accepts switching rules with which $\mathcal{T}_1 \oplus \mathcal{T}_2$ satisfies ψ. Formally, $\mathcal{A}_{\forall\psi}^{\mathcal{T}_1, \mathcal{T}_2} = \langle \{1, 2\}, \Sigma_I, S_1 \times S_2 \times \{1, 2\} \times Q, s_0, \delta', S_1 \times S_2 \times \{1, 2\} \times \alpha \rangle$, where s_0 is a new state and for $who \in \{1, 2\}$ we have $\delta(s_0, who) = \langle \theta_1(who), \theta_2(who), who, q_0 \rangle$ and for all $\langle s_1, s_2, k, q \rangle \in S_1 \times S_2 \times \{1, 2\} \times Q$ we have

$$\text{①}\ \delta'(\langle s_1, s_2, k, q \rangle, who) = \begin{bmatrix} \bigwedge_{i\in\Sigma_I} (i, \langle \eta_1(s_1, i), s_2, who, \delta(q, \langle i, L_k(s_k)\rangle)\rangle) & \text{if } L(q) = 1 \\ \bigwedge_{i\in\Sigma_I} (i, \langle s_1, \eta_2(s_2, i), who, \delta(q, \langle i, L_k(s_k)\rangle)\rangle) & \text{if } L(q) = 2 \end{bmatrix}$$

$\text{●}\ \delta'(\langle s_1, s_2, k, q \rangle, who) = \bigwedge_{i\in\Sigma_I} (i, \langle \eta_1(s_1, i), \eta_2(s_2, i), who, \delta(q, \langle i, L_k(s_k)\rangle)\rangle).$

Intuitively, a state $\langle s_1, s_2, k, q \rangle$ stands for the transducer $\mathcal{T}_1$ being in s_1, the transducer $\mathcal{T}_2$ being in s_2, the transducer that is switched on is $\mathcal{T}_k$, and the automaton $\mathcal{A}_\psi$ is in state q. In the dormant composition, only $\mathcal{T}_k$ changes its state. In both compositions, the O-element of the letter that $\mathcal{A}_\psi$ reads in q is the output of $\mathcal{T}_k$. It is not hard to prove that $\mathcal{A}_{\forall\psi}^{\mathcal{T}_1, \mathcal{T}_2}$ accepts a full tree with directions from Σ_I generated by a transducer $\mathcal{S}$ iff the composition of $\mathcal{T}_1$ and $\mathcal{T}_2$ according to $\mathcal{S}$ satisfies ψ.

We reduced the switching-rule synthesis problem to the nonemptiness problem for $\mathcal{A}_{\forall\psi}^{\mathcal{T}_1, \mathcal{T}_2}$. The number of states of the DPW $\mathcal{A}_\psi$ is doubly-exponential in $|\psi|$, and its index is exponential in $|\psi|$ [20,23]. Therefore, the number of states of the DPT $\mathcal{A}_{\forall\psi}^{\mathcal{T}_1, \mathcal{T}_2}$ is linear in $|\mathcal{T}_1|$ and $|\mathcal{T}_2|$ and doubly-exponential in $|\psi|$, and its index is exponential in $|\psi|$. Since the nonemptiness problem for DPT can be solved in time polynomial in the state space and exponential in the index [5], the upper bound follows. Note that the doubly-exponential complexity is only in terms of $|\psi|$, and the algorithm is polynomial in $|\mathcal{T}_1|$ and $|\mathcal{T}_2|$.

For the lower bound, note that the synthesis problem for LTL is 2EXPTIME-hard already for a formula ψ with $O = \{p\}$. Let $\mathcal{T}_1$ and $\mathcal{T}_2$ be single-state transducers that satisfy p and $\neg p$, respectively. A switching rule for $\mathcal{T}_1$ and $\mathcal{T}_2$ then corresponds to a transducer with $O = \{p\}$, and the synthesis problem for ψ can be reduced to the switching-rule synthesis problem for $\mathcal{T}_1, \mathcal{T}_2$, and ψ.   □

Note that since the switching rule S reads the inputs to all components, nothing prevents it from naively recomputing the output of the components. The essence of a switching rule, however, is to avoid this computation. For example, in the security-camera network discussed in Section 1, a scheduler that implements the software that detects suspicious behaviors is not of much interest. One way to prevent the switching rule from recomputing the output of the components is to restrict its input. In practice, however, optimizing the switching rule obtained in the construction in Theorem 5 would project out the parts that are not essential for the switching rule.

Assumptions on the switching, and hence restrictions on the synthesized switching rule, can be made by replacing ψ by $\psi_{fair} \to \psi$. The automaton $\mathcal{A}_{\forall\psi}^{T_1,T_2}$ then continues to read the I-component of the alphabet from the directions of the tree, the O-component from the active transducer, and reads who from the input tree.

5   Language Characterization

Recall that a component may be switched on forever. Thus, a required condition for a switched system to satisfy a property is that all the underlying components satisfy it. For some properties, this is also a sufficient condition, giving rise to a simple compositional model-checking procedure for them. In this section we seek a characterization of such properties. We solve the problem for the active composition and leave it open for the dormant composition.

In Section 2.1 we showed that the dormant and active compositions correspond to *shuffle* and *merge* of languages. Let $\oplus \in \{\bullet, \mathbb{O}\}$ be a composition operator. We say that a language L is *closed under* $\oplus$ iff $L \oplus L \subseteq L$. That is, for every $u, v \in L$, we have that $u \oplus v \in L$. For example (recall the table in Figure 1), the language $(0 + 1)^*0^\omega$ is closed under both $\bullet$ and $\mathbb{O}$, the language $(01)^\omega$ is closed under $\bullet$ but not under $\mathbb{O}$, the language $(0^*1)^\omega$ is closed under $\mathbb{O}$ but not under $\bullet$, and the language 0^*10^ω is closed under neither $\bullet$ nor $\mathbb{O}$.

As the examples above demonstrate, a language that is closed under shuffle or merge need not be a safety or a co-safety language. It turns out that an exact characterization of the languages that are closed under shuffle or merge is a challenging combinatorial problem. As described below, we have succeeded to obtain an exact characterization for merge. The problem of an exact characterization for shuffle remains open.

Recall that a language L is closed under merge if for every two words $u, v \in L$, all words obtained by locating in position i the i-th letter in either u or v are in L. This means that each of the requirements imposed by L refers to a precise location (e.g., the 4-th letter is 0), or is an eventuality, in which case the requirement in the scope of the eventuality is a safety property (e.g., eventually always 0). Formally, we characterize closure under merge by means of *regular counting*, defined below.

Definition 2. *A language L is* regular counting *if there are $n, k \in \mathbb{N}$ and functions $f_0 : \{0, \ldots, n-1\} \mapsto 2^\Sigma$ and $f_1, f_2 : \{0, \ldots, k-1\} \mapsto 2^\Sigma$ such that for all $0 \le j \le k-1$, we have $f_2(j) \subseteq f_1(j)$ and $w \in L$ **iff** for all $0 \le j \le n-1$ we have $w[j] \in f_0(j)$ and there is $i \ge n$ such that for all $j \ge n$, if $j < i$, then $w[j] \in f_1(j \bmod k)$, and if $j \ge i$, then $w[j] \in f_2(j \bmod k)$.*

Intuitively, the function f_0 describes how the prefix of length n of all the words in L behaves – each location j in this prefix can take letters from the subset $f_0(j)$ of Σ. After the prefix of length n the words in L behaves in some cyclic manner, for a cycle of length k. For some bounded number of locations, this cyclic behavior is described by f_1 – each location j in this infix can take letters from the subset $f_1(j \bmod k)$ of Σ. Eventually, however, the cyclic behavior is described by f_2, which is more restricted than f_1. It is not hard to see that a language L is safety iff it is regular counting with $f_1 = f_2$.

To understand the notion of regular counting better, we now describe an automata-theoretic characterization of it.

Definition 3. *An automaton* $\mathcal{A} = \langle \Sigma, Q, q^0, \delta, \alpha \rangle$ *is a counting automaton if* $\mathcal{A}$ *is a deterministic co-Büchi automaton (DCW) and* Q *can be partitioned into three disjoints sets* $P = \{p_0, \ldots, p_{n-1}\}$, $S = \{s_0, \ldots, s_{k-1}\}$, *and* $S' = \{s'_0, \ldots, s'_{k-1}\}$ *such that:*

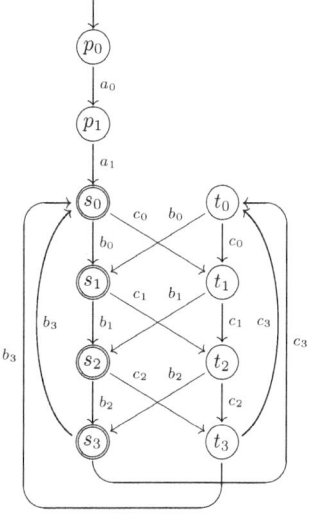

1. *For every* $0 \leq i \leq n-1$, *there is* $\emptyset \neq \Omega_i \subseteq \Sigma$ *such that for all* $\sigma \in \Omega_i$, *we have* $\delta(p_i, \sigma) = p_{i+1}$ *(with* p_n *standing for* s_0) *and for all* $\sigma \notin \Omega_i$, *we have* $\delta(p_i, \sigma) = \emptyset$.
2. *For every* $0 \leq i < k-1$, *there are* $\Omega_i, \Omega'_i \subseteq \Sigma$ *such that* $\Omega_i \cap \Omega'_i = \emptyset$ *and* $\Omega'_i \neq \emptyset$, *such that*
   - *for all* $\sigma \in \Omega_i$, *we have* $\delta(s_i, \sigma) = \delta(s'_i, \sigma) = s_{i+1}$,
   - *for all* $\sigma \in \Omega'_i$, *we have* $\delta(s_i, \sigma) = \delta(s'_i, \sigma) = s'_{i+1}$,
   - *and for all* $\sigma \in \Sigma \setminus (\Omega_i \cup \Omega'_i)$, *we have* $\delta(s_i, \sigma) = \delta(s'_i, \sigma) = \emptyset$.
3. $\alpha = S$.

Example 1. The automaton described in the above figure is a counting automaton accepting the language $a_0 a_1 ((b_0 \vee c_0)(b_1 \vee c_1)(b_2 \vee c_2)(b_3 \vee c_3))^* (c_0 c_1 c_2 c_3)^\omega$.

Proposition 1. *Let* $L \subseteq \Sigma^\omega$. *There exists a counting automaton* $\mathcal{A}$ *such that* $\mathcal{L}(\mathcal{A}) = L$ *if and only if* L *is regular counting.*

We are now ready to state our main theorem for this section.

Theorem 6. $L \subseteq \Sigma^\omega$ *is regular and preserved under merge iff* L *is regular counting.*

The difficult direction is proving that if L is regular and preserved under merge, then L is regular counting. As detailed in the full version, we do this by first proving that if $L \subseteq \Sigma^\omega$ is preserved under merge and is regular, then L is accepted by a deterministic co-Büchi automaton. Essentially, in [10], Landweber proves that a deterministic Rabin automaton has an equivalent deterministic Büchi automaton iff its accepting strongly connected components are upward closed (that is, if S is accepting, so are all components $S' \supseteq S$). We prove that the rejecting strongly connected components of a deterministic Streett automaton for a language L that is preserved under merge are downward closed, and conclude that L can be accepted by a deterministic co-Büchi automaton. We then prove that the states of the deterministic co-Büchi automaton can be partitioned as required in the definition of a counting automaton.

Acknowledgements

We thank Yoad Lustig, Michael Margaliot, Kedar Namjoshi, Dana Porrat, and Gera Weiss for helpful discussions and references.

References

1. Amla, N., Emerson, E.A., Namjoshi, K.S., Trefler, R.J.: Abstract Patterns of Compositional Reasoning. In: Amadio, R.M., Lugiez, D. (eds.) CONCUR 2003. LNCS, vol. 2761, pp. 431–445. Springer, Heidelberg (2003)
2. Antsaklis, P.: Proceedings of the IEEE, special issue on hybrid systems: theory and applications 88(7) (2000)
3. de Jong, H., Gouze, J.-L., Hernandez, C., Page, M., Sari, T., Geiselmann, J.: Qualitative simulation of genetic regulatory networks using piecewise-linear models. Bulletin of Mathematical Biology 66, 301–340 (2004)
4. Emerson, E.: Automata, tableaux, and temporal logics. In: Proc. Workshop on Logic of Programs. LNCS, vol. 193, pp. 79–87. Springer, Heidelberg (1985)
5. Emerson, E., Lei, C.-L.: Efficient model checking in fragments of the propositional μ-calculus. In: Proc. 1st LICS, pp. 267–278 (1986)
6. Erickson, R., Maksimovic, D.: Fundamentals of power electronics. Kluwer, Dordrecht (2001)
7. Hespanha, J., Bohacek, S., Obraczka, K., Lee, J.: Hybrid modeling of TCP congestion control. In: Di Benedetto, M.D., Sangiovanni-Vincentelli, A.L. (eds.) HSCC 2001. LNCS, vol. 2034, pp. 291–304. Springer, Heidelberg (2001)
8. Johansson, K., Lygeros, J., Sastry, S.: Modeling of hybrid systems. Encyclopedia of life support systems (2004)
9. Kozen, D.: Lower bounds for natural proof systems. In: 18th FOCS, pp. 254–266 (1977)
10. Landweber, L.H.: Decision Problems for omega-Automata. Mathematical Systems Theory 3(4), 376–384 (1969)
11. Lafferriere, G., Pappas, G., Sastry, S.: O-minimal hybrid systems. Math. Control Signal Systems 13, 1–21 (2000)
12. Liberzon, D.: Switching in Systems and Control. Birkhauser, Basel (2003)
13. Margaliot, M.: Stability analysis of switched systems using variational principles: an introduction. Automatica 42(12), 2059–2077 (2006)
14. Molisch, A.F., Foerster, J.R., Pendergrass, M.: Channel models for ultrawideband personal area networks. IEEE Wireless Communications 10(6), 14–21 (2003)
15. Mayer, A.J., Stockmeyer, L.J.: The complexity of word problems - this time with interleaving. Information and Computation 115, 293–311 (1994)
16. Pnueli, A.: In transition from global to modular temporal reasoning about programs. In: Logics and Models of Concurrent Systems. NATO Advanced Summer Institutes, vol. F-13, pp. 123–144. Springer, Heidelberg (1985)
17. Pnueli, A., Rosner, R.: On the synthesis of a reactive module. In: Proc. 16th POPL, pp. 179–190 (1989)
18. Pnueli, A., Rosner, R.: Distributed Reactive Systems are Hard to Synthesize. In: 31st FOCS, pp. 746–757 (1990)

19. Rabin, M.: Decidability of second order theories and automata on infinite trees. Trans. of the AMS 141, 1–35 (1969)
20. Safra, S.: On the complexity of ω-automata. In: Proc. 29th FOCS, pp. 319–327 (1988)
21. Safra, S.: Exponential determinization for ω-automata with strong-fairness acceptance condition. In: Proc. 24th STOC (1992)
22. Shorten, R., Leith, D., Foy, J., Kilduff, R.: Analysis and design of AIMD congestion control algorithms in communication networks. Automatica 41, 725–730 (2005)
23. Vardi, M., Wolper, P.: Reasoning about infinite computations. I& C 115(1), 1–37 (1994)
24. Weiss, G., Alur, R.: Automata Based Interfaces for Control and Scheduling. In: Bemporad, A., Bicchi, A., Buttazzo, G. (eds.) HSCC 2007. LNCS, vol. 4416, pp. 601–613. Springer, Heidelberg (2007)
25. Winkler, P.: Mathematical Puzzles: A Connoisseur's Collection, pp. 109–111. A K Peters, Wellesley (2004)

Dataflow Analysis for Properties of Aspect Systems

Yevgenia Alperin-Tsimerman and Shmuel Katz

Department of Computer Science
The Technion, Haifa 32000, Israel
{alpery,katz}@cs.technion.ac.il

Abstract. In this work, data and control flow analysis is used to detect aspects that are guaranteed to maintain some classes of linear temporal logic properties. This is, when the aspects are added to a system that satisfied the desired properties, these properties will remain true. Categories of advices (code segments) and introduced methods of aspects are defined. These categories are shown to be effective, in that they both provide real aid in verification of properties, and are automatically detectable using data and control flow. An implemented automatic data and control flow tool is described to detect the category of each advice and introduced method of an aspect. The results of applying the tool to aspect systems are summarized.

1 Introduction

Aspect-oriented programming (AOP) is designed to aid in separating cross-cutting concerns. The main idea of AOP is to add functionality to an existing (*underlying*) program without modifying its code. Treatment of such concerns without aspects either involves scattering code throughout many parts of a system, or tangling code for the relevant concern with code dealing with separate concerns.

The AspectJ language [8] is an aspect-oriented extension for Java that allows defining separate modules called *aspects*. An aspect defines a set of *pointcuts* (predicates describing points during execution - usually method calls) and *advices* (code segments) for execution at those pointcuts. There are several kinds of advices, which specify where the code of the advice executes relative to the event defined by a pointcut (before, after, or around it). *Before* and *after* advices are executed before or after the defined event, respectively. *Around* advice is executed instead of the event, and uses a *proceed* statement to call the original functionality. The aspect can also introduce new methods and fields. An aspect *weaving* binds the aspect with an *underlying* system (the system without aspects) to produce an *augmented* (or *woven*) system.

Aspects may assign values to the program variables, terminate program execution, and so on. This means that aspects can change the underlying program flow and results. In general, even if the underlying program was already verified, the woven program should be verified from the beginning to prove the woven

K. Namjoshi, A. Zeller, and A. Ziv (Eds.): HVC 2009, LNCS 6405, pp. 87–101, 2011.

program correct. Of course, aspects can add new properties to the system, often in the form of new invariants or post-conditions that connect variables of the underlying system to new variables of the aspect. However, there are also many desirable properties of a system that should be *maintained* when aspects are applied. That is, if such a property held in the underlying system, it will continue to hold in the woven system.

In this paper, we describe an effective automatic dataflow detection tool for categories of aspects in AspectJ with implications for verification, and in particular for categories that guarantee maintaining desirable properties. Thus, dataflow techniques usually applied for compiler optimizations, or possibly program slicing, are used to reduce the need for expensive verification of some properties. Families of temporal logic properties, such as safety or liveness, are proven once-and-for-all to be maintained if an advice in the woven system belongs to a certain category.

In addition to automatically detecting some known categories, we define two new ones: *Public-Call-Correcting* aspects that only perform adjustments that could have been performed by users of the system, and *Control-Safe* aspects that do not influence the control of the underlying system after the advice is executed. Public-Call-Correcting aspects could correct the parameters of external method calls, or, as will be described, introduce a new mode of treatment during system setup. We also consider the categories of new provided services introduced by an aspect and, for the first time, which variables are not affected by an advice or an introduced method and their usage. This allows the output of our analysis to be significantly more informative than previous approaches.

In the following section, some related work is discussed. Then, in Section 3 the categories we consider are defined, and, in Section 4, for each its influence on properties is given (with proof outlines for the new categories). In Section 5 the design and information gathering is described for the dataflow tool yielding category detection, where Subsection 5.4 summarizes applications of the tool. In Section 6 the implications of this approach are briefly considered.

2   Related Work

In the article [7] several categories of aspects are defined according to their semantic characteristics. Some general algorithms for detection of some of the categories are described, but are not implemented. In the article [3], the categories are extended, and their characteristics are described in terms of formal operational semantics, again without any tool for detection.

Data-flow for analysis of aspect systems was first used for standard optimization in compilers [13]. Traditional slicing tools for aspects can be seen in [17] and [19]. In the works [12], [16] and [18] dataflow tools for analysis of aspects are suggested. The classifications are different from those we use, and in particular, the implications for correctness or maintenance of properties are not considered. They basically investigate possibilities of reading/writing to common fields by a single method and aspect advice, or whether the method identified by the pointcut will be executed after the advice. In [16] potential interference among aspects is detected.

3 Categories and Their Implications

3.1 Advice Categories

In this paper an aspect is considered as a set of advices and methods. We thus determine the category of each method and advice separately, as opposed to e.g., [7], where the category of an entire aspect is defined as being spectative, regulative, weakly-invasive, or invasive. Adapting the definitions from [7], we have:

Spectative advice does not modify the underlying program's state or control-flow (it only gathers information in local variables of the aspect).

Regulative advice does not modify the underlying program's state, but its control flow may be terminated (returning to a home "rest" state) or delayed.

Weakly-Invasive advice may modify states and control-flow of the underlying program. However, the execution of the underlying program code in the the the woven program must involve only states which are reachable by the unwoven program.

Strongly-Invasive advice - all other advices.

We will detect the first category above, and special cases of the second, third, and forth. In addition, we define the new categories:

Public-Call-Correcting advice has a pointcut describing public methods that are called outside the underlying program (an environment call), and can only modify variables of the underlying program or parameters of the pointcut method by using public methods or by changing public variables.

Control-Safe advice may modify states of the underlying program, but does not modify its control-flow. More precisely, the advice can change the values of a subset of variables V of the underlying program, but the variables in V have no influence on the control flow of the underlying program.

One application of the Public-Call-Correcting category is to change the mode of operation of a program according to some external conditions. For example, consider a program that works with a data base. In general, this program will connect to a remote data base, but for debug mode the programmer may want it to use a local simulation. For this purpose an aspect can be defined that assigns to the data base variable the relevant value (local or remote data base) before the underlying program execution. The program then can work in different modes depending on which, if any, weaving is performed.

The Control-Safe category is very difficult for manual detection since the analyses of all variables that are influenced by the advice weaving is required. But detection of special cases of this category is useful, since its maintained properties (e.g., termination or responsiveness) are very important and difficult to formally check.

3.2   Provided Services of Introduced Methods

An aspect can change the original program by defining an advice, but it also can provide additional services that extend the underlying program, by supplying an *introduced method* - a public method that is defined in the aspect. The method can be defined as a part of the aspect or as an extension to a class of the underlying program.

The introduced method itself does not have direct influence on a program execution unless it is called. Of course, when an aspect with such a method is woven, the advice of that aspect, an external user, or any other aspect could possibly call this method, but there is no obligation to do so. Effectively, the interface of the underlying program is extended by the introduced method. Clearly, private methods of an aspect do not provide additional services to the system, because they can be used only in the aspect itself and they should not be considered as provided services.

An interesting question is the influence of such introduced methods on the underlying program, when and if the call is performed. In this work the aspect categorization includes the investigation of its provided services.

The classification of introduced methods is performed in a similar way to the classification of advices, although there are differences. A spectative introduced method is the same as a spectative advice, but the regulative category is irrelevant. This is because a method does not have a pointcut, and whether the underlying code will execute is determined in an advice using the introduced method, and not in the method itself. Moreover, the invasive category is divided into *Potentially* and *Guaranteed* subcategories. In the potentially invasive category, the introduced method changes its parameters, but whether these parameters are part of the state of the underlying program can only be determined by considering the actual parameters of each activation. In the guaranteed invasive category, the introduced method is guaranteed to change the state of the underlying system.

4   Maintained Properties of Advice Categories

The purpose of **effective** aspect categorization is to facilitate verification of woven programs, where here we concentrate on maintained properties. For each category identified by the tool, classes of linear temporal logic properties are identified that are maintained.

In order to justify which properties are maintained by advice in a given category, a semantic definition of weaving and of the category is needed. In [7] weaving is defined as a transformation on state machine graphs, and the categories are defined as restrictions on the graphs and allowed transformations. In [3], a structured operational semantics is used to define the allowed actions. Other semantic definitions of aspects, such as [6,15] can also be used. Below, the results from previous work are briefly restated, and for new categories, the proofs informally show the relation between traces of the system augmented with an aspect of the category and traces of the underlying system that is assumed to have a relevant linear temporal logic property (where we consider the usual future fragment, without past modalities). The common modal operators G(from now on), F (eventually), and U (until) can be used, but the next-state operator X should not appear in the properties we wish to maintain. Realtime properties depend on external time variable, and also are not in our framework.

4.1   Strongly Invasive Advice

For strongly invasive advices there is no definite group of properties which are maintained after weaving. But we can consider properties that use variables which are not affected by the advice (neither by data nor by control flow modification).

Definitions for Lemma 1
For the advice a and the property p we will define:
V is a set of all variables in the underlying program.
V_a is a set of variables which are modified by a
DV_a is a set of variables which are dependant on variables in V_a (transitive closure is considered, and DV_a includes V_a)
U_p is a set of variables which are used in p

Lemma 1. For an advice a and a property p that does not relate to realtime or next-state properties if $DV_a \cap U_p = \emptyset$ and p holds in the underlying program, then in the woven program the property p also holds (this means the property p is maintained).

Proof. Consider the projection of the system states on $V - DV_a$ (the variables not dependant on variables in V_a). For each trace of the woven program, this projection is identical to a trace of the unwoven program, except for repetition of the states. Notice that $V - DV_a$ is not affected by the advice. This means that even if a different branch is chosen on some conditional statement, the original values of the variables in $V - DV_a$ and the order of the states cannot be interrupted.

The same state can be repeated more than in the original trace as a result of transitions of the advice. And the same state can be repeated less or more than in the original trace if in the woven trace a different branch was chosen on some condition statement.

According to a lemma of Lamport [9], if two traces differ only in the number of repetitions of states, then all non-next state safety and liveness properties are the same in the two traces. $U_p \subseteq V$, since it is a property of the underlying system. One of the conditions of the lemma is $DV_a \cap U_p = \emptyset$, therefore $U_p \subseteq V - DV_a$ and this means that p is maintained.   □

Conclusion. For the advice a and a property p that does not relate to realtime or next-state properties, if $DV_a \cap U_p = \emptyset$ and p does *not* hold in the underlying program, then in the woven program the property p also does not hold.

Proof. The proof is the same as in Lemma 1, when the projection of a trace that contradicts the property p is considered.   □

4.2 Spectative Advice

Spectative advice does not change the underlying program execution, so the projection of the woven system on the state variables of the underlying system has the same traces, except for repetitions of states that represent the execution of the advice. Such advice therefore maintains all linear temporal logic properties of the underlying program which do not relate to realtime or next-state properties (see [7]).

A spectative advice can also add properties to the augmented program which connect variables of the underlying program and variables of the aspect. But a spectative advice cannot add properties which are related only to variables of the underlying program.

4.3 Regulative Advice

A regulative advice can only delay or terminate program execution. Semantically some transitions are disallowed, but the state variables of the underlying program are not affected. As is shown in [7], if an advice is regulative it maintains all safety properties which a spectative aspect maintains.

4.4 Public-Call-Correcting Advice

A Public-Call-Correcting advice just changes parameters of the method that is described by the pointcut, or changes the underlying program state of public variables before/after the call. A necessary condition is that an external user can perform all operations which are performed by the aspect (i.e., there is no usage of private variables or methods in the advice) and the pointcut describes only the external calls of the method.

Lemma 2. If an advice is Public-Call-Correcting, all safety and liveness properties which were true the underlying system without limiting external public calls, and that do not involve assertions about parameters of the actual call of the user, or realtime or next-state properties, will not be influenced by the advice, and will also hold in the augmented system.

Proof. Only the case of an around advice will be discussed here, since before and after advices are special cases of an around advice where there are no statements before/after a proceed statement. Assume that a Public-Call-Correcting advice that should be woven around method f is given.

According to the definition of Public-Call-Correcting advice the advice activation (an occurrence of the pointcut) will be performed only around a public method that is called from outside the underlying system. Consider two traces: t_1 of the woven system with the aspect, and t_2 of the original underlying system. In t_1, a call of f occurs with some actual parameters, but before f is executed the advice begins, in which some public variables are changed, or public methods are activated, and the parameters of f may be changed. Then f is activated, and afterwards some public variables may again be changed. In t_2 there is no weaving, but the user performs the same operations as in the advice (this can be done because the advice is allowed to perform only actions that an external user could also do) and calls the method with the same parameters as the advice did. After completing f, the user does the corresponding actions done in the advice. In both traces the state of the system and parameters of the call are the same before the first statement in the method f. This means that the program execution till the last statement of f will be exactly the same.

The only difference between the two traces is that t_2 does not have a "pointcut" state where f is called with actual parameters different from those actually used when the advice executes f. Notice that the program cannot make assumptions about the values of its public variables before the user call, thus any property (a linear temporal logic assertion) that does not relate to the actual parameters in the pointcut activation of f (that can be different from those that the advice substitutes) has exactly the same projection onto the state variables that appear in the assertion. Therefore, by reasoning as previously, the property must be maintained.   □

Lemma 3. A Public-Call-Correcting advice can add new properties to the underlying system. The added properties can only represent next-state or realtime properties, or be dependent on the variables which are modified by the advice (i.e., $DV_a \cap U_p \neq \emptyset$, using the notation from Lemma 1).

(Proof is omitted for reasons of space.)

4.5   Control-Safe Advice

A Control-Safe advice maintains the control-flow of the underlying program. So the maintained properties are temporal properties which do not relate to the program state (values of the variables), but to the performed operations and their order. In particular, important properties such as termination will be maintained for the program with a woven Control-Safe advice.

Lemma 4. If an advice is control safe, all assertions about performed operations and their order in traces of the underlying system will also hold in the augmented system.

Proof. According to the definition of Control-Safe advice, the advice does not change variables which can influence control flow of the underlying program. Consider the first control statement in the execution of the underlying system that checks a variable v and chooses between continuing with path A or path B. In the unwoven program, say path A is executed. In the augmented program variable v will be exactly as in the unwoven one, since the aspect cannot change v or variables that v depends on, and it is the first control statement, so it will be executed. Because of this in the augmented program the path A will be executed as well. This means that all operations in path A that were performed in the underlying program also will be performed in the woven program till the next control statement. By induction we will receive that all operations of the underlying program will be performed on the augmented one, although their result can be different from the original.   □

4.6   Relations

Now that the maintained properties for each category are investigated, the relation "is a sub-category" can be defined. A is a sub-category of B means that A maintains at least all properties that B maintains (of course the opposite is not necessarily true). Using results from [7], along with the new lemmas in this paper, the hierarchy seen in Figure 1 can be defined. The lower categories maintain wider classes of linear temporal logic formulas.

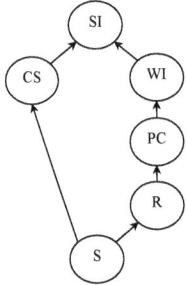

A → B - A is a subcategory of B
S - Spectative advice
R - Regulative advice
PC - Public-Call-Correcting advice
WI - Weakly-Invasive advice
SI - Strongly-Invasive advice
CS - Control-Safe advice

Fig. 1. The hierarchy of subcategories for advices

A similar relation can be defined for the categories of introduced methods: the Spectative category is a sub-category of the Potential-Invasive category and the Potential-Invasive category is a sub-category of the Guaranteed-Invasive category.

5   Detection of the Defined Categories

The categorization tool is built as as extension to the AspectBench Compiler (abc) [2]. It is a complete compiler for the AspectJ language that is built on two extensible frameworks, *Polyglot* and *Soot*. The Polyglot [10] framework is a compiler front-end for the Java programming language, that generates an abstract syntax tree (AST) from the program code while performing the necessary syntax checks. Soot [4] is used by abc as a back-end. It is a bytecode analysis and transformation framework for Java. Soot provides different features which are used for our tool for advices categorization like: Jimple, a typed 3-address intermediate representation that is much easier to analyze than the Java bytecode; several control flow graphs (*CFG*) and services for their construction (e.g., building a CFG for a given method body); different implementations of data-flow analyses (i.e, Forward Flow Analysis); a points-to analysis and so on [4].

The combination of several different analyses for the categorization of the advices allows detecting the category of an advice or introduced method (see Table 1). The categorization is performed on Jimple and it occurs in two phases - before and after the weaving. In practice it is impossible to detect all cases of the defined categories by static analysis. As usual in data flow, the analysis is conservative in that, when an advice is not determined to be in a category, this just means that the advice may not be of that category, while positive detection is certain.

The first step is to check whether there are variables of the underlying program which are modified by an advice (the details of this check are shown in Section 5.1). If there are such variables, then the first conclusion is that the advice is not spectative or regulative. Now a set of variables of the underlying program which are influenced by the advice can be built by slicing (see Section 5.3). If there is no variable used in control flow that is affected by the advice, then the advice is Control-Safe.

If an advice is of the around type, then the proceed call should be analyzed. AspectJ does not limit the number of proceed calls in an advice. Therefore, we need to check whether all paths in the data flow of an advice have exactly one proceed (see Section 5.2). Another check that should be performed for a proceed call is whether it is an *exact proceed* (receives the same argument values as found in the join point and the value returned by proceed must be returned by the advice without modification [18]).

Important cases of regulative advices are detected by a check for synchronization (e.g., a synchronize statement appears in the advice) or program termination (e.g., a System.exit() statement). If an advice throws an exception or contains an infinite loop it can be also considered as a regulative behavior. However, uncaught exception throwing and infinite loops in advices are ignored during the analysis, since such behavior can be considered as a bug in the program, and the program's maintained properties are meaningless in this case.

A check of the pointcut is used to identify Public-Call-Correcting or regulative advices (e.g., if it describes a public method and is external to the system).

The categorization of introduced methods does not require a special analysis, different from the above. All categories can be detected by analysis of variables which are modified in the data-flow of the method.

In this work we have concentrated on the categorization of a single advice or introduced method. But, since we consider all methods and advices of the system at once, interference between advices from different aspects is also treated for one special case: if there is an assignment to a varia ble of aspect B in an advice of aspect A. In this case the category of A is detected relatively to the underlying

Table 1. Checked properties for the categorized advices

Category	Modifies variables of the system	Synchronization or system exit	Exactly one proceed in each path	Exact proceed	Public pointcut
S	Only local for the aspect	Not allowed	Yes	Yes	Not necessary
R1	Only local for the aspect	Allowed	Yes	Yes	Not necessary
R2	Only local for the aspect	Allowed	There is path with no proceed	Yes	Yes
PC	Only public variables	Allowed	Not necessary	Not necessary	Yes
CS	Variables which do not affect program control flow	Not Allowed	Not necessary	Not necessary	Not necessary

S - Spectative advice
R1 - Regulative advice, when the advice contains synchronization or program exit.
R2 - Regulative advice, when the advice cancels execution of the pointcut method.
PC - Public-Call-Correcting advice
CS - Control-Safe advice

```
public class Rational {                      public aspect Reduction {

    private int numerator   = 0;                 Rational around (int x, int y):
    private int denominator = 1;                     (args(x,y) &&
                                                      call(Rational.new(..))) {
    public Rational(int n, int d) {                   Rational f = proceed(x,y);
        setNumerator(n);                              reduce(f);
        setDenominator(d);                            return f;
    }                                             }

    public void setNumerator(int n) {            public void reduce(Rational fr) {
        this.numerator = n;                          int n = fr.getNumerator();
    }                                                int d = fr.getDenominator();
                                                     int tmp = gcd(n, d);
    public void setDenominator(int d) {              fr.setNumerator(n/tmp);
        if (d == 0) throw                            fr.setDenominator(d/tmp);
          new InvalidParameterException();       }

        this.denominator = d;                    public int gcd(int a, int b) {
    }                                                return (b == 0) ? a : gcd(b, a % b);
                                                 }
    // standard getters …                    }
}
```

Fig. 2. Rational Example: class Rational and aspect Reduction

system woven with B and not relatively to the original base system, since the aspect A cannot be added to the underlying program till aspect B is woven and the properties of the pure base system are not relevant already to be maintained.

We will illustrate the analysis steps on a small example (see Figure 2), with a class Rational that describes a fraction, and aspect Reduction that reduces a created fraction just after its construction.

5.1 Summaries

Corresponding to information needed for the columns of Table 1 several data flow analysis should be executed. For each method (or advice) a *modified* set (a set of non-local variables which are modified), flags for existence of synchronization, termination and exception throwing operations in the code and a set of variables which are used in the control flow should be detected.

Initially, for the modified variables detection a bottom-up summary-based algorithm was built. The goal of the algorithm is to build a context-insensitive summary for each method in the augmented program. For this purpose a CFG is built for the underlying system and aspects. Forward Flow Analysis of Soot is used to traverse the graph from the bottom to the top where for each Jimple statement the modified variables are checked. Summaries for each method/advice are combined for the aspect categorization. This architecture is adapted from the work of [16].

In general, a method (or advice) can access its own parameters[1]and local variables, or visible static variables of the system. For each of the above, their instance and static fields must also be considered (recursively). Although all modified variables are checked, we are interested only in changes that can influence the underlying program. So the changes of local variables and their instance fields are ignored in the analysis. The rest of the variable types we will call *dangerous variables*.

The modified set of the current method (or advice) is built by considering assignments and method invocations. For an assignment statement, only if the assigned variable is dangerous is it added to the modified set of the current method (or advice). For a method invocation, the dangerous variables in the modified set of the called method are added to the modified set of the current method (or advice).

To avoid the execution of a full data-flow analysis each time different data need to be gathered, a registration mechanism was defined in the tool. An algorithm for modified variable detection is replaced with a general in-out function that is propagated in the data-flow analysis. For each kind of data that needs to be propagated in the data-flow analysis, actions for a gathering of this data should be added to the in-out general function in a special format.

In the Rational example there are four methods with a nonempty modified set. These are the constructor and setters in the Rational class and the reduce method in the Reduction aspect. The bottom-up analysis finds the modified set of the Rational class functions. *setNumerator* and *setDenominator* change the 'this' object. This means that for the analysis of the *reduce* method the invocation object of *setNumerator* and *setDenominator* - fr - will be modified. fr is a parameter to the function; this means that it is dangerous and will be added to the modified set of the reduce method. The last element analyzed is the advice at the beginning of the Reduction aspect. It invokes the *reduce* method with a local variable as parameter, so the modified set of the advice is empty. During the same analysis we can see that there are no synchronization or termination

---

[1] 'this' is also considered as a parameter.

operations in the advice data-flow. Therefore, this advice seems to be a serious candidate for the Spectative category. The further analyses will show if it is indeed Spectative and will determine the categories of other elements such as the reduce introduced method.

5.2   Exact-Proceed Analysis

The CFG for an advice is built using features of Soot. A DFS algorithm is run on this graph and reports: if there is a path where two or more proceed calls are performed; if there is a path where an exit node was reached, and there was no proceed call, or the standard case where there is exactly one proceed call in each path.

If there is exactly one proceed call in the advice, then we check if it is an exact proceed. For this purpose we need to check that the proceed receives the same arguments as the pointcut method and these variables were not modified in the data flow of the advice from the beginning and until the proceed call. A separate check is made that the advice's returned value (if any) is exactly the same as the one the proceed invocation returns and this value was not modified from the proceed call until the return statement. Each stage is implemented as a DFS on a call graph where each node (statement) is checked for modifying the relevant variables. For the invoked methods their modified set that was calculated by data-flow analysis is checked.

In the Rational example there is one around advice with one path that calls *proceed*. The arguments to the *proceed* are the same as the arguments to the advice and they are not modified till the *proceed* call. But there is a problem with the return value. The variable f stores the value returned by the *proceed*, but it is modified after the *proceed* call. The tool can reveal this by checking the modified set of the *reduce* method, which includes its parameter. This means that the *proceed* of the around advice is not an exact-proceed and the advice has invasive behavior.

The joinpoint of the around advice is the constructor of the Rational class. The constructor is public and is not called in the program itself. The advice itself only modifies public variables, including use of a public method. This means that the advice is Public-Call-Correcting. When an external user calls the constructor, the advice "fixes" the input by replacing the parameters by reduced ones.

5.3   Slicing

General slicing algorithms can be run on the woven system to find the variables that are dependent on the modified set of the advice - *a modified dependent set*. We are interested in both data dependent variables and control dependent variables. Even if the advice is strongly-invasive this information allows identifying the properties that should be verified in the woven system with fuller analyses or testing, since only properties which use variables from the modified dependent set must be rechecked (Lemma 1).

Our intention was to use the slicing algorithms that were developed in Indus [11] on the woven program, although [19] also could be used. The general approach can be optimized by considering only relevant parts of the woven system: the paths where the relevant advice can be applied. Unfortunately, the Indus slicer is difficult to integrate and use as part of a tool chain. Thus, applying a Java bytecode or Jimple formal slicer is left for future work, and is not included in the current tool.

5.4   Experience

Since there is an inclusion relation among the categories, accurate detection means identifying the lowest possible (most specific) category to which it belongs in the hierarchy. In that case, the largest classes of properties are maintained.

For the initial testing of the tool several aspects were defined. For example `SpectativeAdvices` contains spectative advices including cases of callback methods usage, assignments to the fields of the aspect or local advice variables and so on. `InvasiveAdvices` contains strongly-invasive advices including cases of static variables changes, changes of underlying program variables which are visible for the advice, proceed with new (not original) arguments and so on. Separate aspects were defined for introduced methods of different types. JUnit tests were built for consistent checks during development.

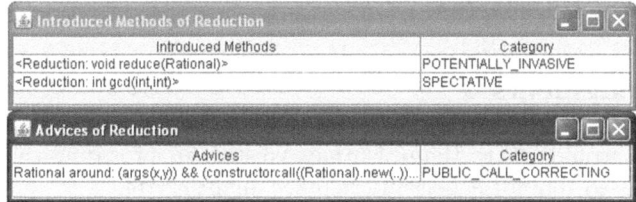

Fig. 3. The output for the example in Figure 2

As a benchmark we took a version of the `HealthWatcher` system that has been used in previous studies [5]. That system has 20 aspects with approximately 45 advices and 30 introduced methods. The benchmark was analyzed manually (by an independent colleague familiar with the definitions) and the category for each advice and introduced method was listed. The tool was run, and the output was compared to the manual results.

Since the tool is under development, only partial matching takes place for the categories. The reasons for this are the presently incomplete treatment of exception handling and aliasing analysis, as well as the treatment of external libraries, that has problematic issues. Overall, for the `HealthWatcher` system: there was accurate detection for 15 of 16 spectative advices, 25 of 26 strongly-invasive advices and 3 of 3 regulative advices.

The tool has GUI output where the list of advices is given and pressing on an advice will lead to the table of introduced methods and advices with appropriate categories (see Figure 3).

A full list of benchmarks including AspectJ benchmarks [1] and AJHotDraw [14], and the results of applying the dataflow tool can be found by clicking on the name of this paper at:
`http://ssdl-linux.cs.technion.ac.il/wiki/index.php/Thesis`

6   Conclusions

In this work we have concentrated on automatically detectable categories of aspects that are useful to show properties are maintained. Therefore there are categories that are not treated here, because their implications for correctness are unclear or nonexistent.

We believe that the tight linking of program properties, proofs of correctness, and static data-flow analysis, as seen in this paper, has significant potential to increase the reliability of systems. The suggested automatic data-flow tool used for identifying categories can increase the understanding of systems with aspects, and more than that it can alleviate part of the need for, e.g., model checking, allowing formal techniques to be applied more selectively. A fuller understanding of the categories of aspects and their implications for correctness can also lead to methodological changes in aspect-oriented software development. For example, aspects in problematic categories should be used sparingly, and be subjected to especially thorough verification.

Acknowledgments

The authors would like to thank Nathan Weston for providing the source code of his abc-based tool for aspect interference analysis [16].

References

1. abc compiler home page, `http://abc.comlab.ox.ac.uk`
2. Allan, C., Avgustinov, P., Christensen, A.S., Dufour, B., Goard, C., Hendren, L., Kuzins, S., Lhoták, J., Lhoták, O., de Moor, O., Sereni, D., Sittampalam, G., Tibble, J., Verbrugge, C.: abc the aspectbench compiler for aspectj a workbench for aspect-oriented programming language and compilers research. In: Companion to the 20th Annual ACM SIGPLAN Conference on Object-oriented Programming, Systems, Languages, and Applications, OOPSLA 2005, pp. 88–89. ACM, New York (2005)
3. Djoko, S.D., Douence, R., Fradet, P.: Aspects preserving properties. In: Proceedings of the 2008 ACM SIGPLAN Symposium on Partial Evaluation and Semantics-based Program Manipulation, PEPM 2008, pp. 135–145. ACM, New York (2008)
4. Einarsson, A., Nielsen, J.D.: A survivors guide to java program analysis with soot (2008)
5. Filho, F., Cacho, N., Figueiredo, E., Garcia, A., Rubira, C.: Exceptions and aspects: the devil is in the details. In: Proceedings of the 14th Intl. Conf. on Foundations of Software Engineering, FSE 2006, pp. 152–162. ACM, New York (2006)

6. Jagadeesan, R., Jeffrey, A., Riely, J.: A calculus of untyped aspect-oriented programs. In: Cardelli, L. (ed.) ECOOP 2003. LNCS, vol. 2743, pp. 54–73. Springer, Heidelberg (2003)
7. Katz, S.: Aspect categories and classes of temporal properties. Transactions on Aspect Oriented Software Development (TAOSD) 1, 106–134 (2006)
8. Kiczales, G., Hilsdale, E., Hugunin, J., Kersten, M., Palm, J., Griswold, W.G.: An overview of AspectJ. In: Lee, S.H. (ed.) ECOOP 2001. LNCS, vol. 2072, pp. 327–353. Springer, Heidelberg (2001), http://aspectj.org
9. Lamport, L.: What good is temporal logic? In: IFIP Congress, pp. 657–668 (1983)
10. Nystrom, N., Clarkson, M.R., Myers, A.C.: Polyglot: An extensible compiler framework for java. In: Hedin, G. (ed.) CC 2003. LNCS, vol. 2622, pp. 138–152. Springer, Heidelberg (2003)
11. Ranganath, V.P., Hatcliff, J.: Slicing concurrent java programs using indus and kaveri. Int. J. Softw. Tools Technol. Transf. 9(5), 489–504 (2007)
12. Rinard, M., Salcianu, A., Bugrara, S.: A classification system and analysis for aspect-oriented programs. SIGSOFT Softw. Eng. Notes 29(6), 147–158 (2004)
13. Sereni, D., de Moor, O.: Static analysis of aspects. In: Proceedings of the 2nd International Conference on Aspect-oriented Software Development, AOSD 2003, pp. 30–39. ACM, New York (2003)
14. van Deursen, A., Marin, M., Moonen, L.: AJHotDraw: A showcase for refactoring to aspects. In: Proceedings AOSD Workshop on Linking Aspect Technology and Evolution, CWI (2005)
15. Wand, M., Kiczales, G., Dutchyn, C.: A semantics for advice and dynamic join points in aspect-oriented programming. ACM Trans. Program. Lang. Syst. 26(5), 890–910 (2004)
16. Weston, N., Taiani, F., Rashid, A.: Interaction analysis for fault-tolerance in aspect-oriented programming. In: Proc. Workshop on Methods, Models, and Tools for Fault Tolerance, MeMoT 2007, pp. 95–102 (2007)
17. Xu, G., Rountev, A.: Ajana: a general framework for source-code-level interprocedural dataflow analysis of aspectj software. In: Proceedings of the 7th International Conference on Aspect-Oriented Software Development, AOSD 2008, pp. 36–47. ACM, New York (2008)
18. Zhang, D.: Aspect impact analysis, Master's thesis, McGill University (August 2008)
19. Zhao, J.: Slicing aspect-oriented software. In: Proceedings of the 10th International Workshop on Program Comprehension, IWPC 2002, Washington, DC, USA, p. 251. IEEE Computer Society, Los Alamitos (2002)

Bisimulation Minimisations for
Boolean Equation Systems

Jeroen J.A. Keiren and Tim A.C. Willemse[*]

Department of Mathematics and Computer Science,
Technische Universiteit Eindhoven,
P.O. Box 513, 5600 MB Eindhoven, The Netherlands

Abstract. *Boolean equation systems (BESs)* have been used to encode several complex verification problems, including model checking and equivalence checking. We introduce the concepts of *strong bisimulation* and *idempotence-identifying bisimulation* for BESs, and we prove that these can be used for minimising BESs prior to solving these. Our results show that large reductions of the BESs may be obtained efficiently. Minimisation is rewarding for BESs with non-trivial alternations: the time required for solving the original BES mostly exceeds the time required for quotienting plus the time for solving the quotient. Furthermore, we provide a verification example that demonstrates that bisimulation minimisation of a process prior to encoding the verification problem on that process as a BES can be arbitrarily less effective than minimising the BES that encodes the verification problem.

1   Introduction

Model checking suffers from the state space explosion problem. Minimising the state space prior to model checking is a well-known strategy for combating this explosion problem, but it is not always obvious that it actually pays to do so in practice. Based on complexity arguments, one can expect that bisimulation minimisation speeds-up model checking for the modal μ-calculus, as in general, the latter requires time exponential in the alternation depth with the size of the state space as root of the exponent (note that some fragments of the μ-calculus can be decided in polynomial time).

   The weakest minimisation that is uniformly permitted in the setting of μ-calculus model checking is a minimisation with respect to strong bisimulation, as the logic can distinguish states up-to bisimilarity. On a case-by-case basis, one can, of course, employ weaker process equivalence relations such as trace equivalence, but judging whether this is the case can require a deep understanding of the system, the formula and process theory. In any case, among all process equivalence relations, bisimulation has the most appealing theoretical time complexity ($\mathcal{O}(m \log n)$ with m the size of the transition relation and n the number

---

[*] This research has been partially funded by the Netherlands Organisation for Scientific Research (NWO) under FOCUS/BRICKS grant number 642.000.602.

K. Namjoshi, A. Zeller, and A. Ziv (Eds.): HVC 2009, LNCS 6405, pp. 102–116, 2011.

of states); in practice, it is rivalled only by some weaker bisimulation relations such as branching bisimulation, with time complexity $\mathcal{O}(mn)$.

The downside of using strong bisimulation for minimising a state space prior to verification is that the minimising capabilities of strong bisimulations are often disappointing. One can improve on this by applying abstractions to the state space prior to applying a bisimulation minimisation, but this suffers from the problems that it (1) requires human intellect, and (2) requires different abstractions each time new properties are verified.

We tackle problems 1 and 2 by employing an intermediate framework, *viz.*, *Boolean equation systems (BESs)*. This framework allows one to encode a variety of verification problems, including the modal μ-calculus model checking problem (see, *e.g.*, [9]). Note that the encoded verification problem can subsequently be answered by computing the solution to the resulting equation system. Solving Boolean equation systems again requires time exponential in the alternation depth of the equation system, with the size of the equation system as root of this exponent (the size of the equation system is polynomial in the size of the state space); efficient algorithms for solving equation systems are based on algorithms for solving *Parity Games*, a framework closely related to BESs.

Instead of performing minimisation of the state space before encoding the verification problem as a Boolean equation system, we apply minimisation techniques on the equation system itself. For these minimisation techniques, we take inspiration from the notion of bisimulation for state spaces. More concretely, we define two notions of bisimulation for Boolean equation systems, *viz.*, *strong bisimulation* and *idempotence-identifying bisimulation*; the latter is tailored specifically to Boolean equation systems and, as far as we are aware, appears to have no natural counterpart in other settings. Both notions are equivalence relations and can be used for quotienting; both are computable in time $\mathcal{O}(m \log n)$, with m the size of all right-hand sides of the equations and n the number of equations. Moreover, strong bisimulation is strictly finer than idempotence-identifying bisimulation, which again is strictly finer than solution equivalence (the latter basically is an equivalence relation based on the local solution of Boolean equation systems, which is typically sufficient for verification). We illustrate that state space minimisation prior to encoding a model checking problem into a Boolean equation system can be arbitrarily less effective than minimising the Boolean equation system that is the result of the encoding.

The advantage of minimisation within the framework of Boolean equation systems is that it does not require human intellect for applying abstractions in order to work. This is because abstraction is taken care of by the encoding to Boolean equation systems. As both bisimulation minimisations respect the solution equivalence, applying minimisations to the Boolean equation system cannot adversely affect the validity of the verification effort, so the approach is fail-safe.

Our bisimulation minimisation techniques provide essential contributions to the framework of *parameterised Boolean equation systems (PBESs)* [5]. The latter are basically high-level, symbolic descriptions of Boolean equation systems,

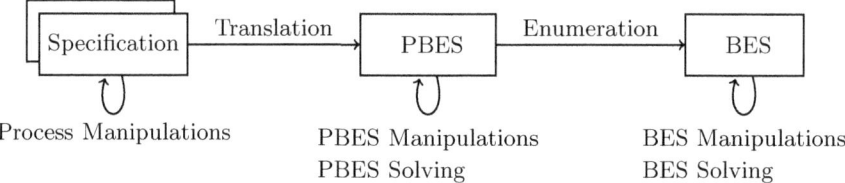

Fig. 1. Verification approach using parameterised Boolean equation systems

which are obtained automatically through encodings of (a variety of) verification problems using symbolic system descriptions as input, see Fig. 1. Various solution strategies for PBESs have been described, among which one finds reductions of PBESs to BESs (see *e.g.* [10], indicated by the arrow linking PBES and BES in Fig. 1). Efficiently minimising the size of the resulting BESs, prior to solving them, is desirable. Observe that the PBES approach to verification avoids generating state spaces altogether; consequently, state space minimisation is not an option in the first place.

We demonstrate the practical value of our approach using a series of experiments that are set in the PBES framework. We rely on state-of-the-art algorithms for solving BESs. The results of these experiments show that indeed, the minimisation of the Boolean equation systems prior to solving is highly rewarding in general: the time required for minimisation is mostly significantly smaller than the time required for solving the unreduced system. The experiments show that in most cases, strong bisimulation and idempotence-identifying bisimulation do not yield significant differences in minimisation capabilities, but, given the more pleasing characteristics of idempotence-identifying bisimulation, the latter is favoured.

Related work. The use of minimisation techniques in combination with verification has been studied in various settings, with mixed results. For LTL verification, Fisler and Vardi [2] show that the total time spent on minimising and verifying exceeds the verification time of the original state space. This can be explained in part, because of the on-the-fly nature of LTL model checking, which does not always require a full construction of a state space. In contrast, in a probabilistic setting, Katoen *et al.* [7] demonstrate that, like in our setting, minimisation mostly pays. Minimisation techniques for Boolean equation systems have received little, if any, attention. In the setting of Parity Games, one finds at least the notion of strong bisimulation and several weaker simulation variants, see [4], but no comparable notion such as idempotence-identifying bisimulation. To the best of our knowledge, in the latter setting, no practical experiments have been conducted using such equivalence relations.

Structure. Section 2 introduces the framework of equation systems, and in Section 3, we define the notions of strong bisimulation and idempotence-identifying bisimulation. In Section 4, we describe a selection of the experiments we conducted using our minimisation methods. The contributions of this paper, and future work are summarised in Section 5. Full proofs, additional lemmata and a report on our >300 experiments can be found in [8].

2 Preliminaries

Boolean equation systems are basically finite sequences of least and greatest fixed point equations, where each right-hand side of an equation is a proposition in positive form. For an excellent treatment of the associated theory, we refer to [9]; in the remainder of this section, we focus on the theory required for understanding the results obtained in this paper.

Definition 1. *A Boolean equation system (BES) $\mathcal{E}$ is defined by the following grammar:*

$$\mathcal{E} \quad ::= \epsilon \mid (\sigma X = f) \, \mathcal{E}$$
$$f, g ::= c \mid X \mid f \vee g \mid f \wedge g$$

where ϵ is the empty BES, $\sigma \in \{\mu, \nu\}$ is a fixed point symbol, X is a proposition variable taken from some set $\mathcal{X}$, f and g are proposition formulae and $c \in \{\mathsf{true}, \mathsf{false}\}$.

For any equation system $\mathcal{E}$, the set of *bound proposition variables*, $\mathsf{bnd}(\mathcal{E})$, is the set of variables occurring at the left-hand side of some equation in $\mathcal{E}$. The set of *occurring proposition variables*, $\mathsf{occ}(\mathcal{E})$, is the set of variables occurring at the right-hand side of some equation in $\mathcal{E}$; for a specific equation we write $\mathsf{rhs}(X)$ to indicate the set of proposition variables occurring in X's equation.

$$\mathsf{bnd}(\epsilon) \stackrel{\Delta}{=} \emptyset \qquad\qquad \mathsf{bnd}((\sigma X = f)\,\mathcal{E}) \stackrel{\Delta}{=} \mathsf{bnd}(\mathcal{E}) \cup \{X\}$$

$$\mathsf{occ}(\epsilon) \stackrel{\Delta}{=} \emptyset \qquad\qquad \mathsf{occ}((\sigma X = f)\,\mathcal{E}) \stackrel{\Delta}{=} \mathsf{occ}(\mathcal{E}) \cup \mathsf{occ}(f)$$

where $\mathsf{occ}(f)$ is defined inductively as follows:

$$\mathsf{occ}(c) \stackrel{\Delta}{=} \emptyset \qquad\qquad\qquad \mathsf{occ}(X) \stackrel{\Delta}{=} \{X\}$$

$$\mathsf{occ}(f \vee g) \stackrel{\Delta}{=} \mathsf{occ}(f) \cup \mathsf{occ}(g) \qquad\qquad \mathsf{occ}(f \wedge g) \stackrel{\Delta}{=} \mathsf{occ}(f) \cup \mathsf{occ}(g)$$

For an equation $\sigma X = f$, we set $\mathsf{rhs}(X) \stackrel{\Delta}{=} \mathsf{occ}(f)$. As usual, for reasons of consistency, we consider only equation systems $\mathcal{E}$ in which every proposition variable occurs at the left-hand side of at most one equation of $\mathcal{E}$. We define an ordering $\lhd$ on bound variables of an equation system $\mathcal{E}$, where $X \lhd X'$ indicates that the equation for X precedes the equation for X'. We say an equation system $\mathcal{E}$ is *closed* whenever $\mathsf{occ}(\mathcal{E}) \subseteq \mathsf{bnd}(\mathcal{E})$. Throughout this paper, we are only concerned with closed equation systems.

  Proposition formulae are interpreted in a context of an *environment* $\eta : \mathcal{X} \to \mathbb{B}$. For an arbitrary environment η, we write $\eta[X := b]$ for the environment η in which the proposition variable X has Boolean value b (note that, for brevity, we do not formally distinguish between a semantic Boolean value and its representation by true and false; likewise, for the operands $\wedge$ and $\vee$).

Definition 2. *Let* $\eta: \mathcal{X} \to \mathbb{B}$ *be an environment. The* interpretation $[\![f]\!]\eta$ *maps a proposition formula* f *to* true *or* false:

$$[\![c]\!]\eta \overset{\Delta}{=} c \qquad\qquad\qquad [\![X]\!]\eta \overset{\Delta}{=} \eta(X)$$

$$[\![f \vee g]\!]\eta \overset{\Delta}{=} [\![f]\!]\eta \vee [\![g]\!]\eta \qquad\qquad [\![f \wedge g]\!]\eta \overset{\Delta}{=} [\![f]\!]\eta \wedge [\![g]\!]\eta$$

The solution *of a BES, given an environment* η*, is inductively defined as follows:*

$$[\![\epsilon]\!]\eta \qquad\qquad \overset{\Delta}{=} \eta$$

$$[\![(\sigma X = f)\ \mathcal{E}]\!]\eta \overset{\Delta}{=} \begin{cases} [\![\mathcal{E}]\!](\eta[X := [\![f]\!]([\![\mathcal{E}]\!]\eta[X := \mathsf{false}])]) & \text{if } \sigma = \mu \\ [\![\mathcal{E}]\!](\eta[X := [\![f]\!]([\![\mathcal{E}]\!]\eta[X := \mathsf{true}])]) & \text{if } \sigma = \nu \end{cases}$$

We refer to computing $[\![\mathcal{E}]\!]\eta$ as *solving* $\mathcal{E}$. Closed equation systems enjoy the property that the solution to the equation system is independent of the environment in which it is defined, *i.e.*, for all environments η, η', we have $[\![\mathcal{E}]\!]\eta(X) = [\![\mathcal{E}]\!]\eta'(X)$ for all $X \in \mathsf{bnd}(\mathcal{E})$. For this reason, we henceforth omit the environment in our considerations and we write $[\![\mathcal{E}]\!]$, and $[\![\mathcal{E}]\!](X)$ instead.

The disjunctions and conjunctions in a proposition formula satisfy the standard rules of logic. For instance, both are semantically idempotent, commutative and associative. This observation justifies the use of a a slightly different grammar for our proposition formulae, introduced next.

Definition 3. *Let* $\mathcal{E}$ *be an equation system. We say that* $\mathcal{E}$ *is in* standard recursive form *(SRF) if the right-hand sides* f *of every one of its equations can be written using the following grammar:*

$$f ::= X \mid \bigvee F \mid \bigwedge F, \qquad \text{where } F \subseteq \mathcal{X}, \text{ with } |F| > 0.$$

where the interpretation is given by the following rules:

$$[\![X]\!]\eta \overset{\Delta}{=} \eta(X) \quad [\![\bigvee F]\!]\eta \overset{\Delta}{=} \bigvee\{\eta(X) \mid X \in F\} \quad [\![\bigwedge F]\!]\eta \overset{\Delta}{=} \bigwedge\{\eta(X) \mid X \in F\}$$

The function $\mathsf{op}(X)$ *for a given equation* $(\sigma X = f)$ *in SRF, returns whether* f *is conjunctive* $(\wedge)$*, disjunctive* $(\vee)$ *or neither* $(\perp)$*.*

Let $\mathcal{B}$ denote the set of all closed equation systems in SRF; in this paper, we are only concerned with equation systems in $\mathcal{B}$. Note that every closed equation system $\mathcal{E}$ can be rewritten to an equation system $\tilde{\mathcal{E}} \in \mathcal{B}$ such that $[\![\mathcal{E}]\!](X) = [\![\tilde{\mathcal{E}}]\!](X)$ for all $X \in \mathsf{bnd}(\mathcal{E})$, *i.e.*, the transformation to SRF preserves the solution of bound variables. This transformation leads to a polynomial blow-up of the original equation system. In [12], the theory outlined in this paper is generalised to arbitrary closed equation systems. An important observation there is that restricting to equation systems in SRF is actually beneficial to the minimisation techniques discussed in this paper.

The *alternation hierarchy* of an equation system, and the derived notion of the *rank* of an equation, can be thought of as the number of syntactic alternations of fixed point signs occurring in the equation system. Note that the alternation hierarchy is not the same as the *alternation depth*: the latter is a measure for the complexity of an equation system, measuring the degree of mutual alternating dependencies, and can be smaller than the alternation hierarchy; it is, however, harder to define and compute.

Definition 4. *Let $\mathcal{E}$ be an arbitrary equation system. The* rank *of some $X \in$ bnd$(\mathcal{E})$, denoted* rank(X), *is defined as* rank$(X) = $ rank$_{\nu,X}(\mathcal{E})$, *where* rank$_{\nu,X}(\mathcal{E})$ *is defined inductively as follows:*

$$\mathsf{rank}_{\sigma,X}(\epsilon) = 0$$

$$\mathsf{rank}_{\sigma,X}((\sigma'Y = f)\mathcal{E}) = \begin{cases} 0 & \text{if } \sigma = \sigma' \text{ and } X = Y \\ \mathsf{rank}_{\sigma,X}(\mathcal{E}) & \text{if } \sigma = \sigma' \text{ and } X \neq Y \\ 1 + \mathsf{rank}_{\sigma',X}((\sigma'Y = f)\mathcal{E}) & \text{if } \sigma \neq \sigma' \end{cases}$$

The alternation hierarchy ah$(\mathcal{E})$ *is the difference between the maximum and the minimum of the ranks of the equations of $\mathcal{E}$. Observe that* rank(X) *is odd iff X is defined in a least fixed-point equation.*

A visual representation of an equation system $\mathcal{E} \in \mathcal{B}$ is its *dependency graph* $\mathcal{G}_{\mathcal{E}}$. This is basically a directed graph with decorated states. Dependency graphs can be used to give a more operational viewpoint of the concept of solution for equation systems, facilitating many of the proofs for theorems appearing in this paper, see [8].

Definition 5. *A dependency graph $\mathcal{G}_{\mathcal{E}}$ of an equation system $\mathcal{E}$ in SRF is a structure $\langle V, \rightarrow, r, l \rangle$, where:*

- *$V = $ bnd$(\mathcal{E})$ is the set of states;*
- *$\rightarrow \subseteq V \times V$ is the transition relation, defined as $X \rightarrow Y$ iff $Y \in$ rhs(X);*
- *$r{:}V \rightarrow \mathbb{N}$ is the rank function, defined as $r(X) = $ rank(X);*
- *$l{:}V \rightarrow \{\wedge, \vee, \perp\}$ is the logic function, defined as $l(X) = $ op(X).*

3 Equivalences

The alternation hierarchy, the number of equations and the complexity of the right-hand sides of these equations account for the computational complexity of the solution for an equation system. Efficient techniques for reducing one or more of these is of the utmost importance. An important step in this direction is to consider equivalence relations for equation systems. An obvious equivalence relation on equation systems is based on the concept of solution for an equation system.

Definition 6. *Let $\mathcal{E}, \mathcal{E}' \in \mathcal{B}$. We say equations for X and Y are solution equivalent, denoted $X \equiv Y$, if $[\![\mathcal{E}]\!](X) = [\![\mathcal{E}']\!](Y)$; we say $\mathcal{E}$ and $\mathcal{E}'$ are solution equivalent, denoted $\mathcal{E} \equiv \mathcal{E}'$, if their first equations are solution equivalent.*

In a possible lattice of equivalence relations on equation systems, $\equiv$ is the coarsest equivalence relation of interest. Deciding $\equiv$ is in NP∩co-NP. Let $\mathcal{E}$ be an equation system. We abbreviate $\mathsf{ah}(\mathcal{E})$ by d, the number of equations in $\mathcal{E}$ by n, the cumulative size of the right-hand sides in $\mathcal{E}$ by m and the size of $\mathcal{E}$ is $m + n$. Algorithms for computing the solution (and thereby deciding $\equiv$) are, *e.g.*, *Small Progress Measures* [6] which runs in $\mathcal{O}(dm(\frac{n}{d})^{\lceil d/2 \rceil})$, *bigstep* [14] which runs in $\mathcal{O}(mn^{\frac{1}{3}d})$, and *Gauß Elimination* [9] which runs in $\mathcal{O}(2^{m+n})$.

In the remainder of this section, we define and study two finer equivalences, *viz.*, *strong bisimilarity* and *idempotence-identifying bisimilarity*, the latter being a subtle adaptation of bisimilarity for equation systems which has, to the best of our knowledge, no natural counterpart in other domains.

3.1   Strong Bisimilarity

Strong bisimilarity (hereafter referred to as *bisimilarity*) for equation systems is inspired by the corresponding notion in domains such as *process theory* and *modal logic*. While bisimilarity has never been defined for equation systems, it is somehow known in the related framework of *Parity Games*, see [4].

Definition 7. *Let $\mathcal{E}, \mathcal{E}' \in \mathcal{B}$. A relation $R \subseteq \mathsf{bnd}(\mathcal{E}) \times \mathsf{bnd}(\mathcal{E}')$ is said to be a bisimulation if, whenever $X \, R \, Y$, then:*

- $\mathsf{rank}(X) = \mathsf{rank}(Y)$;
- $\mathsf{op}(X) = \mathsf{op}(Y)$;
- *for all $U \in \mathsf{occ}(X)$, there is a $V \in \mathsf{occ}(Y)$, such that $U \, R \, V$;*
- *for all $V \in \mathsf{occ}(Y)$, there is a $U \in \mathsf{occ}(X)$, such that $U \, R \, V$.*

We say equations for X and Y are bisimilar, *denoted $X \sim Y$, if there exists a bisimulation relation R such that $X \, R \, Y$; we say $\mathcal{E}$ and $\mathcal{E}'$ are* bisimilar, *denoted $\mathcal{E} \sim \mathcal{E}'$, if their first equations are bisimilar.*

Proposition 1. *Bisimilarity is an equivalence relation over $\mathcal{B}$.*

Let $\mathsf{rhs}(X)_{/R} = \{[X']_{/R} \mid X' \in \mathsf{rhs}(X)\}$ denote the set of classes $[X']_{/R}$ in the right hand side of X with respect to a relation R. Note that bisimilarity $\sim$ is the union of all bisimulation relations, and, as such, is again a bisimulation relation. Bisimilarity can be used to minimise an equation system via *quotienting*. This is achieved by constructing an equation for each equivalence class, using both the rank and the logical operand of the equivalence class as building blocks. Observe that each pair of bisimilar equations $\sigma X = f$ and $\sigma' X' = f'$ satisfies $\mathsf{rank}(X) = \mathsf{rank}(X')$ and $\mathsf{op}(f) = \mathsf{op}(f')$.

Definition 8. *Let $\mathcal{E} \in \mathcal{B}$. The* quotient *of $\mathcal{E}$, denoted $\mathcal{E}_{/\sim}$ is an equation system consisting of equations $\sigma_i C_i = f_i$, for $i \in [1..n]$, where:*

- $C_i \in \mathsf{bnd}(\mathcal{E})_{/\sim}$, *i.e., each $C_i \subseteq \mathsf{bnd}(\mathcal{E})$ is exactly one equivalence class of $\mathcal{E}$;*
- *Let $X \in C_i$ and set $F \overset{\Delta}{=} \mathsf{rhs}(X)_{/\sim}$.*

- *In case* $\mathsf{op}(X) = \bigwedge$, *set* $f_i \overset{\Delta}{=} \bigwedge F$;
- *In case* $\mathsf{op}(X) = \bigvee$, *set* $f_i \overset{\Delta}{=} \bigvee F$;
- *In case* $\mathsf{op}(X) = \bot$, *set* $f_i \overset{\Delta}{=} C_j$, *where* $C_j \in F$;

- *Order equations such that* $C_i \lhd C_j$ *iff there is some* $X \in C_i$ *such that for all* $X' \in C_j$, $\mathsf{rank}(X) \leqslant \mathsf{rank}(X')$ *and* $X \lhd X'$ *in* $\mathcal{E}$.

Note that the above construction satisfies that $\mathcal{E} \sim \mathcal{E}_{/\sim}$ for arbitrary $\mathcal{E} \in \mathcal{B}$.

Theorem 1. *The relation* $\sim$ *is strictly finer than* $\equiv$.

Note that strictness follows from the fact that $(\nu X = Y)\,(\mu Y = X) \equiv (\nu X' = Y')\,(\nu Y' = X')$, but not $(\nu X = Y)\,(\mu Y = X) \sim (\nu X' = Y')\,(\nu Y' = X')$. As an illustration of our minimisation techniques, consider the following example.

Example 1. Consider the Labelled Transition System given below:

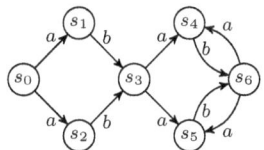

The equation system encoding $s_0 \models \nu X.[a]\mu Y.\langle b \rangle (Y \vee X)$, and its dependency graph are as follows (see, *e.g.*, [9] for the encoding of the modal μ-calculus model checking problem into BES).

$(\nu X_{s_0} = Y_{s_1} \wedge Y_{s_2})$
$(\nu X_{s_3} = Y_{s_4} \wedge Y_{s_5})$
$(\nu X_{s_6} = Y_{s_4} \wedge Y_{s_5})$
$(\mu Y_{s_1} = Y_{s_3} \vee X_{s_3})$
$(\mu Y_{s_2} = Y_{s_3} \vee X_{s_3})$
$(\mu Y_{s_3} = Y_{s_3})$
$(\mu Y_{s_4} = Y_{s_6} \vee X_{s_6})$
$(\mu Y_{s_5} = Y_{s_6} \vee X_{s_6})$
$(\mu Y_{s_6} = Y_{s_6})$

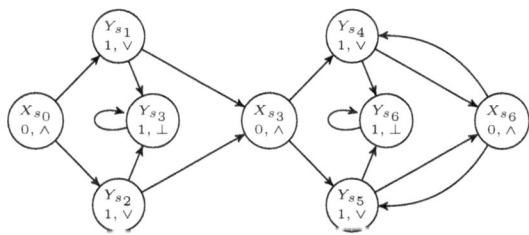

Minimising the equation system module strong bisimulation leads to the following equation system; its associated dependency graph is depicted next to it.

$(\nu X_{s_0} = Y_{s_1} \wedge Y_{s_1})$
$(\mu Y_{s_1} = X_{s_0} \vee Y_{s_3})$
$(\mu Y_{s_3} = Y_{s_3})$

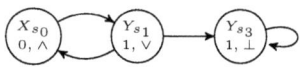

Compared to the original equation system, the minimised equation system is roughly 65% smaller. Observe that in both the original equation system and the reduced equation system, the solution to X_{s_0} is the same, which follows from the fact that both are bisimilar. □

3.2   Idempotence-Identifying Bisimulation

The definition of a quotient for bisimilarity is rather awkward, as equations of the form $\sigma X = X' \wedge X''$, or $\sigma X = X' \vee X''$, with $X' \sim X''$ cannot be minimised further to $\sigma X = X'$ (see also the example in the previous section); doing so nevertheless would change the operand of the equation, leading to a violation of $\mathcal{E} \sim \mathcal{E}_{/\sim}$. From a logical viewpoint, it does not make sense to discriminate between these equations. We observe that a logical operand of an equation is only of importance when it is applied to proposition variables from distinguishable classes; in any other case, the bisimulation relation should be oblivious to the logical operands. These considerations lead us to consider a weaker definition of bisimilarity, called *idempotence-identifying bisimilarity*, which appears to be more natural in the setting of equation systems.

Definition 9. *Let $\mathcal{E}, \mathcal{E}' \in \mathcal{B}$. A relation $R \subseteq \mathsf{bnd}(\mathcal{E}) \times \mathsf{bnd}(\mathcal{E}')$ is said to be an* idempotence-identifying bisimulation *if, whenever $X \, R \, Y$, then:*

- $\mathsf{rank}(X) = \mathsf{rank}(Y)$;
- *if* $\mathsf{op}(X) \neq \mathsf{op}(Y)$ *then for all* $U \in \mathsf{occ}(X)$ *and* $V \in \mathsf{occ}(Y)$: $U \, R \, V$;
- *for each* $U \in \mathsf{occ}(X)$ *there is a* $V \in \mathsf{occ}(Y)$ *such that* $U \, R \, V$;
- *for each* $V \in \mathsf{occ}(Y)$ *there is a* $U \in \mathsf{occ}(X)$ *such that* $U \, R \, V$.

We say equations for X and Y are idempotence-identifying bisimilar, *denoted* $X \sim_{ii} Y$, *if there is an idempotence-identifying bisimulation relation R such that $X \, R \, Y$; we say $\mathcal{E}$ and $\mathcal{E}'$ are* idempotence-identifying bisimilar, *denoted $\mathcal{E} \sim_{ii} \mathcal{E}'$, if their first equations are idempotence-identifying bisimilar.*

Proposition 2. *Idempotence-identifying bisimilarity is an equivalence relation over $\mathcal{B}$.*

Quotienting, based on idempotence-identifying bisimulation, requires a subtle modification of the quotienting for bisimulation. In case we are constructing an equation $\sigma_i C_i = f_i$, where C_i is again the equivalence class of a set of bisimilar equations, f_i is defined as C_j in case $\mathsf{rhs}(X_i)_{/\sim_{ii}} = \{C_j\}$ for all $X_i \in C_i$. In particular, this avoids introducing awkward equations such as $\sigma_i C_i = \bigwedge\{C_j\}$. All other cases are in full agreement with Def. 8. Note that $\mathcal{E} \sim_{ii} \mathcal{E}_{/\sim_{ii}}$.

Theorem 2. *We have:*

1. *the relation $\sim$ is strictly finer than $\sim_{ii}$;*
2. *the relation $\sim_{ii}$ is strictly finer than $\equiv$.*

The following proposition demonstrates that idempotence-identifying bisimilarity and solution equivalence sometimes coincide.

Proposition 3. *Let $\mathcal{E} \in \mathcal{B}$ be of the form $\mathcal{E}_0 \mathcal{E}_1 \mathcal{E}_2$, with $\mathcal{E}_1 \in \mathcal{B}$. Suppose for all $X, X' \in \mathsf{bnd}(\mathcal{E}_1)$, we have $\mathsf{rank}(X) = \mathsf{rank}(X')$. Then $\mathcal{E}_{1/\sim_{ii}} = \mathcal{E}_{1/\equiv}$.*

In words, closed sub-equation systems consisting of equations all of the same rank, can be reduced to a single equation. A special case is when $\mathcal{E}_0 = \mathcal{E}_2 = \epsilon$, in which case the closed equation system $\mathcal{E}_1$ reduces to a single equation. Note that the above result does not hold in the bisimulation setting. In particular, the above result shows that idempotence-identifying bisimilarity can yield a substantially greater reduction, by an arbitrarily large factor, than bisimilarity. The following example illustrates that the same holds when comparing bisimilarity at a process level to idempotence-identifying bisimilarity.

Example 2. Let N be an arbitrary positive number. Consider the process described by the following set of recursive processes (using process algebra style notation):

$$\{S = \sum\{a \cdot X(n) \mid n \leqslant N\}, \quad X(0) = a.X(0) + b.X(0), \quad X(n+1) = b.a.X(n)\}$$

A visualisation of S for $N = 3$ is depicted at the right; S consists of $2(N + 1)$ states, which cannot be minimised further using strong bisimilarity. Define the following μ-calculus formula $\phi \overset{\Delta}{=} \nu Y.\langle a\rangle([a]\text{false} \wedge \nu Z.\langle b\rangle\langle a\rangle Z)$. The equation system $\mathcal{E}$, encoding $S \models \phi$ has $N + 1$ equations. $\mathcal{E}$ is closed and each of its equations has rank 0. Following Proposition 3, quotienting of $\mathcal{E}$ yields the equation system $\nu X = X$, for arbitrary N. Note that one could reduce the labelled transition system (LTS) underlying S with respect to trace equivalence, yielding an LTS of size 2, which, however, no longer satisfies ϕ. This

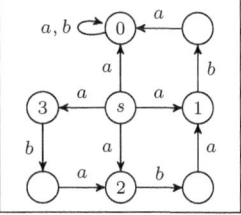

is, of course, in general the case, as no process equivalence weaker than strong bisimilarity preserves the full modal μ-calculus.   □

3.3   Decidability

We can use variations of the well known partition refinement algorithm by Paige and Tarjan [11] for deciding both strong bisimilarity and idempotence-identifying bisimilarity, running in $\mathcal{O}(m \log n)$ time. This algorithm iteratively refines the partitioning given a splitting criterion, where a block is split iff there are two equations that are not equivalent. Equivalence of two equations $\sigma X = f$, $\sigma' X' = f'$ given a current partitioning P is decided by the predicate $\mathsf{Eq}_{\sim}(\mathsf{P}, \mathsf{X}, \mathsf{X}')$ for strong bisimulation and $\mathsf{Eq}_{\sim_{ii}}(\mathsf{P}, \mathsf{X}, \mathsf{X}')$ for idempotence-identifying bisimulation, where Eq is defined as follows.

$$\mathsf{Eq}_{\sim}(\mathsf{P}, \mathsf{X}, \mathsf{X}') \overset{\Delta}{=} \text{rank}(X) = \text{rank}(X') \wedge \text{rhs}(X)/_P = \text{rhs}(X')/_P$$
$$\wedge \ \text{op}(f) = \text{op}(f')$$
$$\mathsf{Eq}_{\sim_{ii}}(\mathsf{P}, \mathsf{X}, \mathsf{X}') \overset{\Delta}{=} \text{rank}(X) = \text{rank}(X') \wedge \text{rhs}(X)/_P = \text{rhs}(X')/_P$$
$$\wedge \ (|\text{rhs}(X)/_P| = 1 \vee \text{op}(f) = \text{op}(f'))$$

That is, in the case of strong bisimulation two equations are equivalent if their ranks and Boolean operators match, and their right-hand sides contain the same classes, whereas for idempotence-identifying bisimulation the requirement for having the same Boolean operators is lifted in case both equations have only one equivalence class in their right-hand sides.

4   Experiments

To test the effectiveness of the two minimisations introduced in the previous section, intended mainly to increase efficiency of solving equation systems resulting from typical verification problems, we ran a large set of verification experiments consisting of model checking and process equivalence checking problems. We present a representative selection of the experiments that we have carried out; an exhaustive listing of our results can be found in [8].

Setup. All experiments were run on a workstation consisting of 8 Dual Core[1] AMD Opteron(tm) Processors running at 2.6Ghz, with 128Gb of shared main memory, running a 64-bit Linux distribution using kernel version 2.6.24. We adapted an off-the-shelf, competitive C implementation by Blom and Orzan [13] for computing the bisimulation minimisations for LTSs, such that bisimulation and idempotence-identifying bisimulation for BESs can be computed efficiently.

The BESs were solved using a development version of PGSolver tool [3].[2] The timings we report have been obtained using the *bigstep* [14] algorithm, enhanced with a set of heuristics for speeding-up the algorithm, as well as without these enhancements. Note that bigstep outperformed the Small Progress Measures algorithm in all our experiments; both are state-of-the-art algorithms for Parity Games. All BESs were generated from *parameterised Boolean equation systems* using the mCRL2 tool suite[3], without generating state spaces first.

4.1   Process Equivalence Experiments

We consider the problem of deciding branching bisimilarity between two processes, encoded as PBES [1]. As input to the equivalence checking problems, we used four descriptions of well-studied communications protocols, *viz.*, the one-place buffer (OPB), two variations of the Alternating Bit Protocol (ABP) and the Concurrent Alternating Bit Protocol (CABP). For each protocol, we varied the size of the set of messages M that could be exchanged from $|M| = 1, 2, 4, 8, 16, 32$. Table 1 shows (1) the size of the original BESs in SRF, and (2) the size after reduction using $\sim$ and $\sim_{ii}$. Note that both reductions are capable of eliminating the dependency on $|M|$. The speedups in solving the BESs that we observed are comparable to the results of Section 4.2, and omitted for brevity.

4.2   Model Checking Experiments

A second batch of experiments is conducted using problems stemming from the μ-calculus model checking problem. We report on our experiments using two complex communications protocols, *viz.*, the Onebit Protocol (OP) and a Sliding Window Protocol with window size 2 (SWP_2). We check for the validity

---

[1] Note that none of our experiments employ dual-core features.
[2] Obtained from Oliver Friedmann through private communication.
[3] See `http://www.mcrl2.org`, revision 6175 (release branch).

Table 1. Sizes of BESs encoding the branching bisimulation verification problem, before and after applying bisimulation minimisations

BES Size Statistics

BES	Size before reduction for $|M| = 1, 2, 4, 8, 16, 32$						Size after reduction	
	1	2	4	8	16	32	$\sim$	$\sim_{ii}$
ABP$_1$ - ABP$_2$	12,193	24,711	50,755	106,875	235,243	556,491	1,462	1,460
ABP$_1$ - CABP	109,706	238,418	553,394	1,413,554	4,054,706	13,020,338	21,329	21,311
ABP$_1$ - OPB	366	840	2,136	6,120	19,656	69,000	75	74
ABP$_2$ - CABP	148,082	320,378	738,410	1,868,234	5,302,922	16,872,458	21,329	21,311
ABP$_2$ - OPB	482	1,115	2,867	8,315	26,987	95,435	75	74
CABP - OPB	4,922	12,018	33,266	103,986	358,322	1,318,578	1,253	1,253

of five modal formulae of increasing complexity. Solving times for OP are summarised in Figure 2, for SWP$_2$ in Figure 3. For all protocols, we verified absence of (I) deadlock and (II) livelock, and the possibility to infinitely often (III) receive a certain message, (IV) receive all messages and (V) receive some message if it is infinitely often enabled. Statistics about the reductions in size modulo strong bisimulation and idempotence-identifying bisimulation are summarised in Table 2.

Table 2. Sizes of BESs encoding the model checking problems, before and after applying bisimulation minimisations

BES Size Statistics

Model	Property	Size before reduction for $|M| = 2, 3, 4, 5$				Size after reduction	
		2	3	4	5	$\sim$	$\sim_{ii}$
OP	I	578,050	2,083,394	5,417,986	-	5	2
	II	1,100,802	3,933,506	10,172,418	-	8	7
	III	619,010	2,179,826	5,603,586	-	26,171	26,171
	IV	1,238,023	6,539,482	22,414,349	-	26,174	26,173
	V	3,358,981	17,664,522	60,344,071	-	71,206	71,205
SWP$_2$	I	71,090	269,282	728,386	1,614,602	5	2
	II	139,698	525,026	1,413,954	3,125,802	2,268	7
	III	80,146	291,842	773,890	1,695,082	9,097	9,097
	IV	160,295	875,530	3,095,565	8,475,416	9,100	9,099
	V	456,037	2,520,258	8,957,959	24,599,908	25,354	25,353

4.3 Discussion

Both strong bisimulation and idempotence-identifying bisimulation minimisation show a significant reduction in the size of the BESs. In all cases minimising the BES modulo either of the equivalences and solving the reduced BES outperforms solving the original BES. In addition, reducing modulo strong bisimulation is slightly faster than reducing modulo idempotence-identifying bisimulation in our prototype. This is expected to be a consequence of the additional checks that need to be carried out in an implementation of idempotence-identifying bisimulation, combined with the small difference in size between the systems reduced modulo strong bisimulation and idempotence-identifying bisimulation.

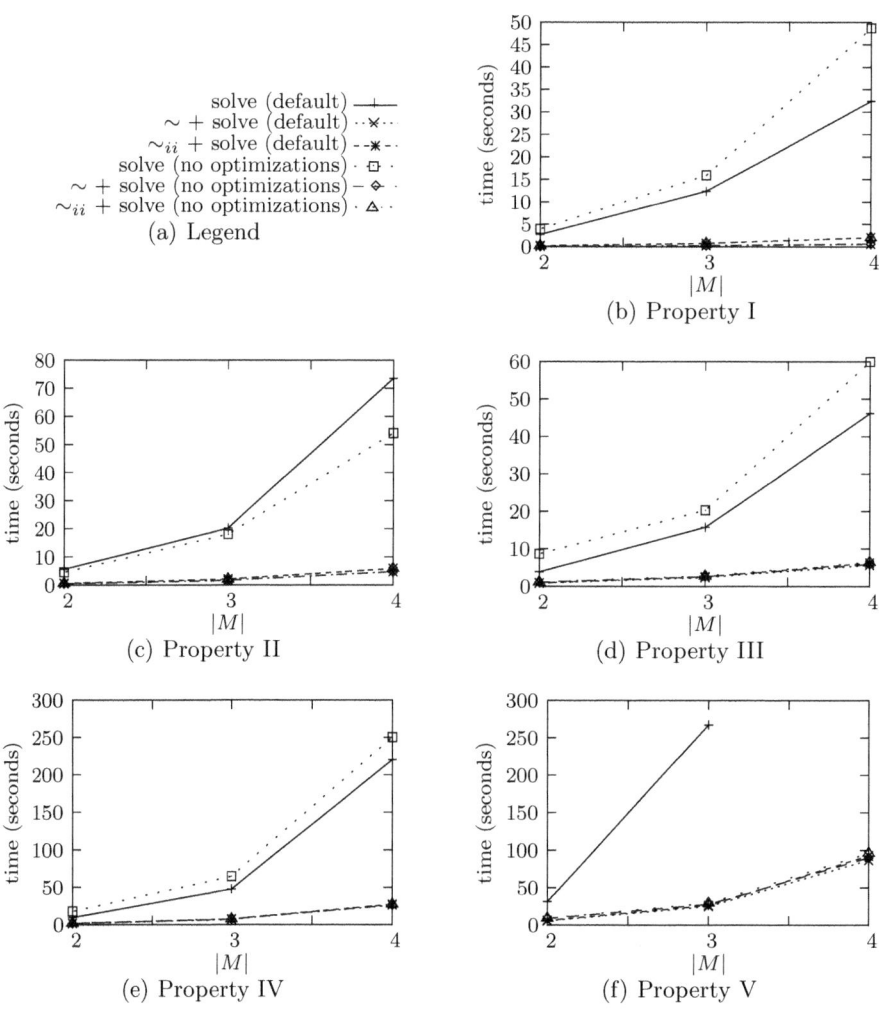

Fig. 2. Timing results for OP; time-out is set to 300 seconds

The graphs of Figures 2 and 3 clearly demonstrate the effect of using bisimulation minimisation prior to solving a BES. They also show the effect of enabling optimisations in the PGSolver, although the effect is in much less dramatic than the effect of our minimisations. Based on our experiments we believe that in practice a bisimulation reduction should be performed prior to solving the BES.

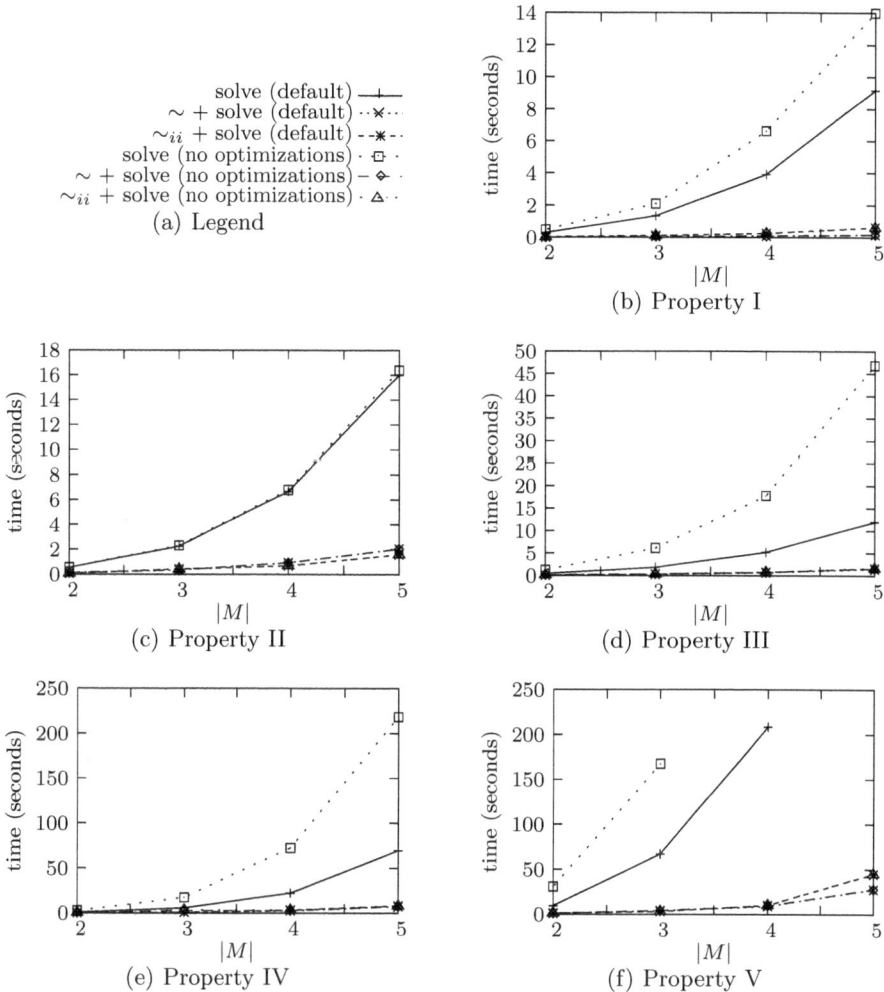

Fig. 3. Timing results for SWP with buffer size 2; time-out is set to 300 seconds

5   Conclusions

In this paper, we have defined two equivalence relations for BESs, *viz.*, *strong bisimilarity* and *idempotence-identifying bisimilarity*. The former takes inspiration from the definition of bisimilarity in settings such as process theory and logic. The latter is a modification of strong bisimilarity, leading to an equivalence that is more natural in the setting of equation systems.

Experiments using a prototype implementation for minimising with respect to our two types of bisimulation indeed confirm that enormous reductions are quite commonplace. Moreover, our measurements show that it pays to minimise before solving: the time required for minimising is more than made up for by the time gained in solving the minimised equation system.

Several topics remain to be investigated. Among these is the investigation of weaker equivalence relations for equation systems. Here, *stuttering equivalence* serves as a source of inspiration, both because of its attractive computational complexity, and because of its capability of achieving far greater minimisations than strong bisimilarity. A related topic that we are currently pursuing is the development of proof theory for *parameterised Boolean equation systems* [5] based on bisimilarity. We are convinced this will lead to more concise proofs in this setting.

References

1. Chen, T., Ploeger, B., van de Pol, J., Willemse, T.A.C.: Equivalence checking for infinite systems using Parameterized Boolean Equation Systems. In: Caires, L., Li, L. (eds.) CONCUR 2007. LNCS, vol. 4703, pp. 120–135. Springer, Heidelberg (2007)
2. Fisler, K., Vardi, M.Y.: Bisimulation minimization and symbolic model checking. Formal Methods in System Design 21(1), 39–78 (2002)
3. Friedmann, O., Lange, M.: Solving parity games in practice. In: Liu, Z., Ravn, A.P. (eds.) ATVA 2009. LNCS, vol. 5799, pp. 182–196. Springer, Heidelberg (2009)
4. Fritz, C., Wilke, T.: Simulation relations for alternating parity automata and parity games. In: Ibarra, O.H., Dang, Z. (eds.) DLT 2006. LNCS, vol. 4036, pp. 59–70. Springer, Heidelberg (2006)
5. Groote, J.F., Willemse, T.A.C.: Parameterised Boolean Equation Systems. Theor. Comput. Sci. 343(3), 332–369 (2005)
6. Jurdziński, M.: Small progress measures for solving parity games. In: Reichel, H., Tison, S. (eds.) STACS 2000. LNCS, vol. 1770, pp. 290–301. Springer, Heidelberg (2000)
7. Katoen, J.-P., Kemna, T., Zapreev, I.S., Jansen, D.N.: Bisimulation minimisation mostly speeds up probabilistic model checking. In: Grumberg, O., Huth, M. (eds.) TACAS 2007. LNCS, vol. 4424, pp. 87–101. Springer, Heidelberg (2007)
8. Keiren, J., Willemse, T.A.C.: Bisimulation minimisations for boolean equation systems. Technical Report CSR09-17, Eindhoven University of Technology (2009)
9. Mader, A.: Verification of Modal Properties Using Boolean Equation Systems. PhD thesis, Technische Universität München (1997)
10. Mateescu, R., Thivolle, D.: A model checking language for concurrent value-passing systems. In: Cuellar, J., Sere, K. (eds.) FM 2008. LNCS, vol. 5014, pp. 148–164. Springer, Heidelberg (2008)
11. Paige, R., Tarjan, R.E.: Three partition refinement algorithms. SIAM J. Comput. 16(6) (1987)
12. Reniers, M.A., Willemse, T.A.C.: Analysis of Boolean equation systems through structure graphs. In: Proc. of SOS 2009. EPTCS (2009) (to appear)
13. Blom, S., Orzan, S.: Distributed state space minimization. Int. J. STTT 7(3) (2005)
14. Schewe, S.: Solving parity games in big steps. In: Arvind, V., Prasad, S. (eds.) FSTTCS 2007. LNCS, vol. 4855, pp. 449–460. Springer, Heidelberg (2007)

Synthesizing Solutions to the Leader Election Problem Using Model Checking and Genetic Programming

Gal Katz and Doron Peled

Department of Computer Science
Bar Ilan University
Ramat Gan 52900, Israel

Abstract. In recent papers [13,14,15], we demonstrated a methodology for developing correct-by-design programs from temporal logic specification using genetic programming. Model checking the temporal specification is used to calculate the fitness function for candidate solutions, which directs the search from initial randomly generated programs towards correct solutions. This method was successfully demonstrated by constructing solutions for the mutual exclusion problem; later, we also imposed some realistic constraints on access to variables. While the results were encouraging for using the genetic synthesis method, the mutual exclusion example includes some limitations that fit well with the constraints of model checking: the goal was finding a fixed finite state program, and its state space was moderately small. Here, in a more realistic setting, we challenge the problem of synthesizing a solution for the well known "leader election" problem; under this problem, a circular, unidirectional network with message passing is seeking the identity of a process with a maximal value. This identity, once found, can be used for synchronization, breaking symmetry and other network applications. The problem is challenging since it is parametric, and the state space of the solutions grows up exponentially with the number of processes.

1 Introduction

Automatic synthesis of correct software is a very difficult problem. Pnueli and Rosner [28] showed that the construction of a distributed implementation from linear temporal logic specification is, in general, undecidable; with some constraints on the architecture, the problem becomes highly intractable: nonelementary for a pipeline architecture or one directional ring. Genetic programming (GP) is a directed search in the space of syntactically fitting programs for a correct instance in an attempt to synthesizing correct code. Rather than being a completely comprehensive search, it allows progressing from one version of the code to another by means of small changes (mutations). It employs some random process and the ability to combine portions of code from different versions together (crossover). It does not commit to a certain path of search for a long time by abandoning the current search and restarting, and by searching simultaneously from different points.

K. Namjoshi, A. Zeller, and A. Ziv (Eds.): HVC 2009, LNCS 6405, pp. 117–132, 2011.

In [13,14,15] we presented an approach for the genetic construction of programs given an LTL specification and some structural constraints on the programs (the allowed constructs, the set of variables and their intended use, the concurrent architecture). This method was exercised to generate existing and new solutions for the mutual exclusion problem. This was done by developing a tool based on our ideas of combining model checking and GP. The tool was fed with the LTL specification for mutual exclusion and with some structural restrictions (e.g., the set of variables that can be used), and automatically generated programs satisfying those properties.

This classical synchronization problem is a good example of a complicated task that programmers may find difficult to deal with. Mutual exclusion with two processes is also a problem that fits well with the natural restrictions of model checking: it is a finite state problem, with a fixed number of processes, and the state space of the solutions is rather small. Still, a comprehensive search for correct solutions takes considerable amount of time [2].

In this paper we want to tackle a more challenging problem, that of finding a solution to the leader election in a circular (i.e., ring) unidirectional network of message passing. Under this problem, the processes in the network are collaborating to find the process with the maximal id value. This will allow assigning that process a special role in future network interactions (e.g., initializing protocols, breaking ties in votes). This protocol poses several challenges, stretching the boundaries of using model checking as a method for directing the genetic process.

Parametric solution. The problem of leader election is parametric, and should work for any number of processes in a ring. Parametric model checking was studied extensively. In general, this is shown to be an undecidable problem [1]. Model checking solutions often include the use of an induction on the number of processes [19,32], and involve a manual step of applying the induction. Our solution involves iterating the model checking on an increased number of processes, depending on the fitness of the solution. This means that for processes that obtain a higher fitness value, showing a better potential for being in a proximity of the desired solution, we perform the model checking with a higher number of processes. This does not guarantee that a given solution works with any number of processes, and a manual check is still useful after the synthesis is completed. However, the solution is at least guaranteed to work with up to a large number of checked processes, hence, being a good candidate for being a parametric solution.

Exponential blowup of the state space. Performing model checking on the known solutions for the leader election problem, e.g., as it is done in the standard examples of SPIN [10], reveals, not surprisingly, that there is an exponential explosion of the state space with the number of processes. Partial order reduction [11] can be a big help in taming this explosion. We provide a theoretical a-priori analysis of the leader election problem, and show that it belongs to a family of problems that utilizes a considerable amount of partial order reduction. We use this observation in our synthesis. On the other hand, there is another source of state space explosion; the leader election also belongs to a family of problems where the executions highly depends on the initial data, i.e., the unique ids of the processes (we refer here only to leader election algorithms that are based on comparisons between the ids of the processes). A similar difficulty with the complexity of model checking, being related to the many cases of values initially distributed among processes, appears for example in model checking of sorting algorithms.

Quantitative preference. One of the important criteria for accepting a solution of the leader election problem is based on the number of messages sent. It was for many years an open problem whether it is possible to find a solution where the number of messages sent is in the worst case $O(n \log n)$ rather than $O(n^2)$, where n is the number of processes. Since the solution is parametric, we compare message growth with the number of processes as part of the fitness of the solution. Moreover, there is a known lower bound of $\Omega(n \log n)$ messages (see [26]). These measures are used not only to direct the search towards an efficient solution, but also to prevent it from solutions that are specific to some particular sizes of ring.

The rest of the paper is organized as follows. Section 2 gives background on the leader election problem, and on our previously used combination of model checking and genetic programming. Section 3 describers the new ideas and enhancements that we use for the leader election problem. In Section 4, the experimental results are shown, and Section 5 concludes the paper.

2 Preliminaries

2.1 The Leader Election Problem

The problem of leader election appeared originally in token ring protocols. It is needed, e.g., when a token circulates in a unidirectional ring of processes with asynchronous message passing, allowing its owner to initiate various protocols. However, sometimes the token is lost, requiring the participating processes to perform some algorithm for obtaining exactly one leader that will finally have the token. In the leader election algorithms, each process initially owns a unique id (normally, a small integer). By sending messages, the processes along the ring decide on one of the processes to be the leader, commonly the process that initially has the maximal id. We assume that there are no failures in delivering messages, and that no processes permanently fail during the algorithm.

A solution with $O(n^2)$ messages was given by Le Lann [21], and optimized by Chang and Roberts [3]. A solution with $O(n \log n)$ messages in the worst case was given by Dolev, Klawe and Rodeh [5], as well as by Peterson [26].

2.2 Model Checking Based Genetic Programming

Genetic Programming [18] is a method for the automatic synthesis of computer programs by an evolutionary process. It searches through the state space of well formed programs, syntactically restricted by the problem statement (e.g., having a certain number of variables or communication patterns). The search starts with some randomly generated programs, and progresses, guided by calculating a fitness value for the generated candidates. The fitness value is used for selecting candidates that will be mutated or merged between one step of the search and the next one. The fitness function is usually calculated based on some given scenarios or test cases.

The programs we generate are represented using syntactic trees, where each node represents a syntactic object, i.e., a segment of the program, a constant or a variable. This allows performing the different mutations on the nodes: Deletion, Insertion, Replacement and Reduction [13]. These mutations may require completing the code into a syntactically correct program. For example, when a node representing a loop is created, a subtree that includes the loop condition and the loop body must also be generated.

The crossover operation allows picking up sections from several candidates and gluing them together, again maintaining the syntactic correctness. In our system, we currently do not use the crossover operation. While being a powerful operator, it is sometimes debatable in the GP community, and some works indicate that good results can be achieved even when using only the mutation operation [4].

The GP algorithm we use in this work progresses through the following steps. We first create an initial population of candidate solutions. Then we randomly choose a subset of μ candidates. We create new λ candidates by applying *mutation* (and, optionally, *crossover*) on the above μ candidates. The fitness function for each of these candidates is calculated, and only μ candidates from the combined set of the λ new candidate and the original μ candidates are selected proportionally to their fitness. The selected candidates then replace the μ candidates originally selected. This is repeated until a perfect candidate is found, or until the maximal permitted number of iterations is reached.

Johnson [12] suggested the use of temporal logic for calculating the fitness, based on the number of properties that hold for the candidate solution. We suggested the use of fine-grained model checking analysis [13,14,15], which does not only allow checking whether temporal properties hold or fail, but provide some intermediate possibilities.

The temporal properties are specified using Linear Temporal Logic (LTL) [27]. The syntax of LTL is defined over a finite set of propositions $\mathcal{P}$, with typical element $p \in \mathcal{P}$, as follows:

$$\varphi ::= true | p | \varphi \vee \varphi | \neg \varphi | X \varphi | \varphi \, U \, \varphi \tag{1}$$

Let M be a finite structure $(S, s_0, E, \mathcal{P}, L)$ with states S, an initial state $s_0 \in S$, edges $E \subseteq S \times S$, a set of propositions $\mathcal{P}$, and a labeling function $L : S \mapsto 2^{\mathcal{P}}$. For simplicity, we assume that each state in S has a successor. This can be forced by adding to each state without successors a self loop, marked with a special symbol ϵ. A *path* in S is a finite or infinite sequence $\langle g_0 g_1 g_2 \ldots \rangle$, where $g_0 \in S$ and for each $i \geq 0$, $g_i E g_{i+1}$. An *execution* is an infinite path, starting with $g_0 = s_0$. Sometimes executions are further restricted to satisfy various fairness assumptions.

We denote the ith state of a path π by π_i, and the suffix of π from the ith state by π^i. The LTL semantics is defined for a suffix of an execution π of M as follows:

$\pi \models true$.

$\pi \models p$ if $p \in L(\pi_0)$.

$\pi \models \varphi_1 \vee \varphi_2$ if either $\pi \models \varphi_1$ or $\pi \models \varphi_2$.

$\pi \models \neg\varphi$ if it is not the case that $\pi \models \varphi$.

$\pi \models X\varphi$ if $\pi^1 \models \varphi$.

$\pi \models \varphi U \eta$ if there exists some i such that $\pi^i \models \eta$ and for each $0 \leq j < i$, $\pi^j \models \varphi$.

We say that a structure M (or the corresponding program that is modeled as M) satisfies φ if for each execution π of M, $\pi \models \varphi$. We use the logical connections to define additional temporal operators, e.g., $\varphi_1 \rightarrow \varphi_2 = (\neg\varphi_1) \vee \varphi_2$, $\varphi_1 \wedge \varphi_2 = \neg((\neg\varphi_1) \vee (\neg\varphi_2))$, $\Diamond\varphi = true U\varphi$, $\Box\varphi = \neg\Diamond\neg\varphi$, etc.

The fitness calculation is based on the following ideas. According to LTL, a property φ has a universal flavor: a program satisfies a property if all of its executions satisfy it. Traditional LTL model checking [22,31] translates the negation of the specification property, $\neg\varphi$, into a Büchi automaton or a similar structure. The intersection of this automaton and one that represents the state space can be searched (e.g. with a variant of Depth First Search [9]) to look for a path representing an execution of the program satisfying $\neg\varphi$. If the intersection is empty, the property φ is satisfied by the candidate program. This gives the highest fitness level with respect to φ (denoted *level 3*). If this does not hold, it is possible to check whether *none* of the executions, or *some* of them, do satisfy φ. This is done by translating φ into an automaton and checking the intersection as above. Then the program is assigned lower fitness levels *0* or *1* respectively.

Another intermediate level of satisfaction (*level 2*) is considered. Accordingly, although there are some executions that do not satisfy the given property, each prefix of such an execution can be completed into an execution of the candidate program that does satisfy the property. In [13], a dedicated algorithm for this level of satisfaction was used. In [23], a formalism that takes an LTL specification and use it for quantifying over paths was presented. This can, in particular, capture all the levels of satisfactions described here. A generic model checking algorithm for this formalism was given. The intuition behind that algorithm is that one may need to know at various points in the execution whether the execution so far can be extended into one which satisfies, from its beginning, the property φ or $\neg\varphi$. To do that, one needs to record at each point of the execution the various states of the Büchi automaton representing the checked property φ that are currently possible. If this automaton is nondeterministic, there can be more than one such state. This makes the analysis exponentially more difficult than simple model checking, in the size of the checked property, and the complexity grows from PSPACE to EXSPACE. Fortunately, this growth is only in the size of the LTL specification, which is typically rather small, and not in the size of the checked system.

In addition to the correctness criteria required by the programming problem, there can be some *quality* criteria. This can involve the size of the program, and limitations on the number of access to variables (e.g., number of accesses before entering a critical section), or messages sent, as will be demonstrated later.

The requirements from the program are not only restricted to linear temporal logic formulas. In fact, there are also some important structural restrictions. This includes, for example, the number of variables per each process, and their use (shared read or write by other processes), and the communication structure.

3   Generating Parametric Programs

In previous papers, we described the basic method of combining genetic programming and model checking, and used the problem of mutual exclusion between two processes as a case study. After rediscovering the classical known solutions [13], we added some more practical structural requirements on the use of variables, and discovered small improvement to current, non-trivial solutions [14]. Using two-process mutual exclusion as a target for code synthesis enjoys several advantages. It is a well defined problem, with a standard specification. The state space of this problem is not very large. This is important for our method and tool, since model checking is performed on many (thousands or more) candidates, not just one program. Finally, the constructed program is a finite state program. This is the case where model checking is decidable, and, combined with the small state space, even efficient. The deeper model checking analysis has a higher complexity than standard model checking in terms of the checked property, but since we use rather small specification, this is still manageable.

Encouraged by the results of our experiments, we were motivated to apply our method to some more challenging synthesis problems, e.g., parametric ones. Then the finite state assumption needs to be abandoned. Although one may sometimes view a computer as a finite state system, this is not the standard way to think about algorithms. Numerical algorithms may perform on a computer with a bounded word size, but they are designed to work also on computers with any word size. Moreover, this finite state assumption does not hold when considering algorithms that work on constantly changing network architecture (such as the Internet), or algorithms that work on arrays of various sizes, such as sorting. For these cases, the finite state view can help verify just an *instance* of the problem, certainly not the complete algorithm. The problem of verification of infinite state space is in general undecidable. So is the problem of verifying parametric systems, as shown by Apt and Kozen [1] for cyclic parametric networks, and for the unidirectional cycle by Suzuki [30]. There are a lot of model checking techniques for infinite state systems. But only very few of them provide a complete automatic decision procedure for some well defined cases; these includes, most notably, a model checking decision procedure for pushdown automata and for Presburger arithmetic [24]. Other analytical methods, such as abstraction and induction can be quite effective in many cases, but often require human intervention.

For the case of unidirectional rings, Emerson and Namjoshi [7] showed how, for properties that are symmetric in a fixed number of processes, the parametrized model checking problem can be reduced to that of checking only a small number of symmetric processes. Emerson and Kahlon [6] suggested a useful framework for parametrized model checking of rings with tokens passing, allowing changes to the tokens values. They proved that when some limitations are imposed on the tokens and possible transitions, the problem can be reduced to the verification of a small number of processes. The framework was used for verifying the Le Lann Leader election protocol [21], but the limitations rule out the verification of more advanced leader election algorithms, such as Chang and Roberts' improvement [3] or Peterson's algorithm [26].

The challenge we took is to synthesize solutions for the problem of leader election in a ring topology. This is a parametric problem: the number of processes is not fixed. Moreover, the state space of this problem grows exponentially with the number of processes involved. In addition to the state space explosion due to concurrency, there is an additional complication of a state space explosion due to the data. Even if the given values that the processors posses are restricted to be between 1 and n, the number of processes, there are still $(n - 1)!$ possibilities to allocate them in a ring. The structure of this problem is thus related to other problems that are parametric in nature, such as concurrent sorting.

3.1   State Space Reduction

State space explosion is a central problem for model checking. There are a large number of suggested solutions. However, the problem of model checking is PSPACE-complete in both the size of the LTL specification and the size of the code [29]. Many model checking algorithms that attempt to combat the state space explosion work well in practice in many cases, but do not guarantee to work efficiently for every case. The problem of leader election may give rise to an exponential growth of the state space in the number of processes. The state space explosion is even more problematic for our goal of using model checking as a procedure for genetic programming; model checking is then performed on tens of thousands of candidate solutions.

Model checking of the solution to the leader election in [5] can be done with SPIN [10]. One example is included with the standard distribution of the tool. SPIN performs this algorithm quite efficiently with its built-in partial order reduction, described in [11]. The growth in space is linear in this case. Without partial order reduction, the state space grows exponentially with the number of processes. SPIN distribution includes other two examples that share the same behavior: the Sieve of Eratosthenes, for calculating concurrently prime numbers and concurrent sorting.

The reason for this optimal behavior of the partial order reduction is that these problems have a special structure. The processes run in a distributed way, without shared variables, but with message passing. In addition, the only non-determinism is resulted from concurrency. The communication structure is fixed as a pipeline, and messages progress in one direction, without the possibility of overtaking messages. The idea behind the partial order reduction is that many executions are equivalent up to commuting the order of independent (concurrent) transitions; when the specification cannot distinguish between such executions, it is sufficient to check representatives for all the equivalence classes. Under the above restrictions for a given distribution of ids, there is, in fact, only one equivalence class per initial state.

In our model, there are only transitions local to a process and asynchronous (buffered) communication between adjacent processes of the ring in one direction. Specifically, a local transition can be commuted with any other transition of other processes. Two communication transitions that do not involve the same process, can also be commuted. Perhaps less obviously, but as in the theory of conditional independence [16], and as used in partial order reduction [11,8], a send and a receive involving the same queue can also be commuted. When both a *send* and a *receive* of the same queue are enabled (when the message queue is neither empty, nor full), we will prefer to execute the *receive* first, resulting in smaller buffers. The scheduling we will use for model checking will thus give priority to executing the local transitions until there is only a communication transition enabled.

Based on this analysis, we implemented in our tool a technique that is related to the partial order reduction algorithms by Overman [25] and Lamport [20]. In generating the state space for the leader election problem, a communication is followed by a maximal finite sequence of enabled transitions together as a large atomic action, and preferring a *receive* action on a *send* from the same buffer.

The above analysis is done for a particular set of initial distribution of ids to the different processes, or, equivalently, different initial states. Note that the SPIN example for leader election actually runs with only one initial assignment of values. In practice, there is an exponential number of ways to distribute n numbers into n processes on a ring (in fact, exactly $(n-1)!$ ways). We can use one representative execution sequence per each initial distribution. However, the partial order reduction does not assist in reducing the complexity due to the initial distribution of values. The same observation applies to the concurrent sorting example, which is checked for a particular arrangement of values.

There are several subtle points that need to be mentioned. The syntactic restriction of the generated candidates is a part of our methodology. Synthesizing code for the mutual exclusion problem, which we performed in [13,14], was restricted to programs with shared variables, and in fact to a particular set of variables. In the leader election case, the coordination between processes is through message passing. The fact that it is enough to check only one representative execution sequence per initial distribution of ids, and that either, all such equivalent executions satisfy the specification, or all of them do not satisfy it, also means that one of the fitness levels that we use for ranking the candidate solutions (*level 2* described above) is redundant: there are no prefixes of a bad execution that branch into a good execution. This is a bit unfortunate, since the finer analysis was shown to be useful for the convergence of the genetic process. On the other hand, *level 2* was the reason that the model checking analysis was in EXPSPACE in the size of the LTL specification [23], rather than PSPACE, as in classical model checking [29]. Thus, when removing this level, we make the model checking more efficient.

3.2   Convergence for Parametric Problems

Perhaps the main problem in model checking based genetic programming is to make the process converge into an acceptable solution. The genetic programming process is probabilistic; it involves generating some random instances, and also the possibility of abandoning the solution and starting from scratch. However, if this process is repeated many times without convergence, we are seeking to change the parameters of the fitness function that directs the search. At this point of the research, we are interested in fine tuning the search through changing various parameters. This includes adding specification properties; even if they are implied by already existing specification. This can help in generating more intermediate levels of fitness and help towards convergence. Another possibility is changing the distribution of the fitness levels between properties and levels of satisfaction (when the property is satisfied by some executions but not all).

In the leader election problem, the convergence was experienced to be more difficult than in our previous experiments with synthesizing mutual exclusion protocols. The problem is, by definition, parametric. We do not have any completely automatic procedure to check for correctness with respect to any number of processes. The model checking of a specific configuration (i.e., number of processes, and the order of their ids), uses a specification with only a few properties. In fact, since our syntax allows processes to announce leadership, but not to retract it, the following single LTL property will suffice:

$$\Diamond \Box (\text{NumberOfLeaders} = 1)$$

While this may be desirable from the verification point of view, it poses a major difficulty for a gradual progressing process such as genetic programming. Therefore, our first task was adding more properties, even if they can be implied by the above single property. An example for that, is splitting the property into two properties a safety property asserting that the number of leaders never exceeds one, and a liveness property assuring a leader will eventually be elected (see Table 1). This can guide the search into programs satisfying only one of the properties at a first stage, and then progressing into a solution satisfying both.

One of the methods for achieving gradual improvement that we have used in [13,14] was the ranking of properties by different levels of fitness. As shown in the previous section, our leader election programs are treated as having only a single execution per each initial distribution of ids, which turns the above analysis to be irrelevant, and hence removes the useful fitness level 2, which we used in previous work. However, another source for multiple executions in our case stems from the initial possible distributions of ids between the processes. Even when restricting ourselves to a fixed number of n processes, a complete verification must check all of the $(n-1)!$ possible permutations of initial ids. Machine Learning techniques, and GP usually cope with such parametric problems by selecting (explicitly, or randomly) a set of configurations which are used as test cases. The main risk with this strategy is over-fitting, i.e., a convergence into solutions that perfectly satisfy all of the test cases, but are not general, and therefore, do not satisfy other unchecked cases. Our experiments for the leader election problem indeed showed that when checking only several permutations of ids, the search has frequently led into specific, non-general solutions. This was true even when more than a half of the permutations were checked.

Therefore, we check *all* of the possible permutations for a fixed number of processes. Due to the fast model checking technique used (only one execution path is checked for each nodes permutation), this could be easily done even for six processes, where the number of permutations is $(6 - 1)! = 120$. Yet, this does not mean that programs at early stages of the search must satisfy all of the permutations at once. One common technique is to use the number of satisfied permutations as a part of the fitness score. However, we chose to use a more qualitative measure by considering only three cases:

- *None* of the permutations satisfies the property,
- *Some* of the permutations satisfy the property, but others do not,
- The property is satisfied by *all* of the permutations.

These fitness levels are in some sense identical to three out of the four fitness levels we have used in [13] although in our case, the source for various executions is the initial configuration, rather than the nondeterminism of processes scheduling.

Even though this technique increased the rate of convergence into good solutions, there were still cases of generated algorithms which behaved perfectly on configurations with up to n processes, but failed on configurations with a larger number of processes. In order to overcome that, we used additional special properties. These properties are not directly implied from the standard leader election properties, but they hold for any parametric solution. Followed are the properties we used:

1. A process cannot declare a leadership without first considering the values of all other processes. It does not mean that the process itself has to make all of the comparisons by itself; it can rely on the information gathered by processes preceding it in the ring. However, the chain of decisions must include all of the processes. Otherwise, it necessarily means that the process ignored at least one process while making its decision. Therefore, if the ignored process has the maximal id, the algorithm was wrong.

   In order to verify this property, we add a *history* local variable to each process, which stores the number of preceding processes in the ring, whose values the process could consider so far. This variable is initially set to zero for each process. When sending a message, the sender attaches its local variable to the sent message, and the receiving process sets its *history* variable to $min(msg.history + 1, n)$, where $msg.history$ is the variable attached to the received message, and n is the number of processes in the ring. Since information can only flow unidirectionally through a set of successive processes, the value of *history* can only be monotonically increased. For our needs, its maximal value can be bounded to n. This variable is a function of the history of the computation, and is not a part of the resulted code. In particular, there are no decisions in the program that depend on the value of this variable. It is merely used for the purpose of verification. Then, if a processes p announces itself as a leader, the following proposition must hold: $p.\text{history} = n$.

2. The lower bound for the number of messages a leader election algorithm must send in the worst case is $\Omega(n \log n)$ (see [26]). While this bound is asymptotic and hence difficult to check explicitly, we can conclude that at least one process must send a non-constant number of messages; otherwise, the total number of message is at most $O(n)$. Since the algorithm has to deal with any number of processes, we deduce that it must contain a loop that contains an instruction for sending a message (if not, the number of sent messages is bounded by the program length).

   This property is somewhat nonstandard. We cannot directly capture this with a particular LTL property. To capture it, we recorded during the model checking search each occurrence of a "Send" transition with its executing process and location counter of that process, ignoring all other state variables. If such a transition is repeated, then at least one process executes a loop with messages sending. Note that this test is a necessary, but not a sufficient condition, since it only verifies that the loop is executed at least twice.

Another classical way of encouraging general solutions is to define a secondary measure that increases the program's score as its size decreases. This measure was successfully used in all of our works, and in many others as well ([17], for instance).

A more advanced measure regards the message complexity. Historically, the worst and average case message complexities of the published leader election algorithms were improved from $O(n^2)$ [21] to match the known lower bound, achieving complexity of $O(n \log n)$ [5,26], and then further improvements were made to the constants in the complexity formulas. The worst case message complexity of our algorithms was calculated by summing the "Send" transitions during the execution path of each permutation, and choosing the biggest sum among them. However, the asymptotically faster algorithms may be actually slower when running on small number of processes. For instance, Peterson's basic algorithm [26] requires $2n \lfloor \log n \rfloor + n$ messages in the worst case, which overtakes Chang and Roberts' algorithm [3] of $\frac{n(n+1)}{2}$ messages only for $n \geq 14$ processes. In order to focus on the general behavior of the algorithms, we calculated the complexity for several successive $n's$, and then used the *difference* between the results as a measure; this is because the rate of growth of the asymptotically better complexity is smaller (which provides a better fitness value) even for a small value of n. This allows to prefer $O(n \log n)$ over $O(n^2)$ algorithms.

4   Experimental Results

We extended our previously developed tool [13,14] in order to support the new features described throughout this article. Table 1 depicts the properties and measure that were checked. The following building blocks were available for the generated programs:

- `UID` - a constant with a different value for each process.
- `R,S,T` - three local variables for each process.
- Assignment instructions.
- Boolean conditions based on comparisons between variables and constants, the `True` and `False` literals, and the `Or` and `And` operators.
- `If` conditions, and `While` loops.
- `Send(val)` - sending the value `val` to the next process in the ring.
- `Receive(T)` - receiving a value from the previous process in the ring into the variable `T`.
- `AnnounceLeader(val)` - allows a process to announce a leadership with the maximal value on the parameter `val`. The process then enters an infinite loop. The announcement increases by one the value of the global variable *NumberOfLeaders*.

Table 1. Leader Election Specification

No.	Definition	Remarks
1	$\Box\neg$ (NumberOfLeaders > 1)	Safety
2	$\Diamond\Box$ (NumberOfLeaders > 0)	Liveness
3	Leader.value $= n$	Leader holds max value
4	Leader.history $= n$	Leader's decision may be sound
5	For $i = 1..n$, $\Diamond\Box$(P_i.buffer is empty)	All message buffers are finally empty
6	Minimize program's size	Secondary measure
7	Minimize worst case complexity	Secondary measure

The tests were run on a PC with an Intel 3GHz processor. At each run, a population size of 100 programs was randomly generated, and evolved until either a perfect solution was found, or a large number of iterations was executed without a fitness improvement. Most of the runs started by finding an idle program that trivially satisfies the safety property 1, or the one-lined program

```
AnnounceLeader(UID)
```

which satisfies property 2. On many runs, the following program was discovered next:

```
Send(UID)
Receive(R)
if (R < UID)
    AnnounceLeader(UID)
```

This is the simplest program during the genetic process, that has discovered the idea of electing a leader by some message passing and comparison. For all permutations, at least one leader is elected (thus fully satisfying property 2), and for most permutations, even more than one. However, since there is a permutation for which exactly one leader is elected (the one where the values of processes are descendingly ordered), the program also partially satisfies property 1.

Many runs successfully converged into perfect solutions. The following solution sends n^2 message in the worst case, as in Le Lann's algorithm [21]. However, this algorithm discovered a technique used by more advanced algorithms, that allows a process to become a relay process by just passing messages.

```
While (True)
    Send(UID)
    Receive(R)
    if (R >= UID)
        While (R != UID)
            Receive(R)
            Send(R)
        AnnounceLeader(UID)
```

Another algorithm was found which is identical to the one of Chang and Roberts [3], and sends $\frac{n(n+1)}{2}$ messages in the worst case:

```
Send(UID)
While (S != UID)
    Receive(S)
    if (S > UID)
        Send(S)
AnnounceLeader(UID)
```

At this point in our experiments, the genetic process has not yet discovered any efficient solution of message complexity $O(n \log n)$.

5   Conclusions

Synthesizing correct-by-design algorithms is a challenging task. The combination of model checking and genetic programming presented in [13,14,15] was shown successful for synthesizing solutions for a problem that is both a classical finite state problem, and of typically small state space. Handling programming problems with a large or infinite state space is difficult for model checking, and even more so for a method that uses model checking as a subroutine that is called thousands of times. Still, from a practical point of view, we wanted to tackle a nontrivial problem that is both parametric and, by being highly concurrent, of state space that is growing exponentially. The classical problem of leader election was chosen to demonstrate our ideas.

We did not find any of the many solutions for model checking of infinite state space sufficiently general to be used as a building block for our tool. Thus, our solution was to apply model checking of instances of the leader election in an incremental way, for a number of processes that grows with the fitness level of the candidate solution. This does not guarantee correctness for an unbounded ring size, but provides a large confidence, when the instance size grows considerably. The state space explosion was handled here using a method that is related to partial order reduction. These solutions are not as generic as we would have liked for the synthesis of large or infinite state space. We consider them as practical building blocks that demonstrate the power of the model checking based genetic programming approach.

For the problem of leader election, we found the problem of converging the genetic search highly challenging. Since the problem is parametric and because of the approach of enlarging the number of processes that we check with the fitness level, we quickly learned that the algorithms constructed tend to be specific for some ring sizes. To generate a solution for unbounded rings, we had to use not only linear temporal logic properties, but additional structural checks based on lower bounds on the number of messages passed by the algorithm.

Experimentally, our tool was able to find some solutions and building blocks for the leader election, in particular, those with $O(n^2)$ messages.

Acknowledgments

We would like to thank Hubert Garavel for suggesting to look at the leader election problem.

References

1. Apt, K.R., Kozen, D.C.: Limits for automatic verification of finite-state concurrent systems. Inf. Process. Lett. 22(6), 307–309 (1986)
2. Bar-David, Y., Taubenfeld, G.: Automatic discovery of mutual exclusion algorithms. In: Fich, F.E. (ed.) DISC 2003. LNCS, vol. 2848, pp. 136–150. Springer, Heidelberg (2003)
3. Chang, E., Roberts, R.: An improved algorithm for decentralized extrema-finding in circular configurations of processes. ACM Commun. 22(5), 281–283 (1979)
4. Chellapilla, K.: Evolving computer programs without subtree crossover. IEEE Trans. Evolutionary Computation 1(3), 209–216 (1997)
5. Dolev, D., Klawe, M.M., Rodeh, M.: An $O(n \log n)$ unidirectional distributed algorithm for extrema finding in a circle. J. Algorithms 3(3), 245–260 (1982)
6. Emerson, E.A., Kahlon, V.: Parameterized model checking of ring-based message passing systems. In: Marcinkowski, J., Tarlecki, A. (eds.) CSL 2004. LNCS, vol. 3210, pp. 325–339. Springer, Heidelberg (2004)
7. Emerson, E.A., Namjoshi, K.S.: On reasoning about rings. Int. J. Found. Comput. Sci. 14(4), 527–550 (2003)
8. Godefroid, P., Pirottin, D.: Refining dependencies improves partial-order verification methods (extended abstract). In: Courcoubetis, C. (ed.) CAV 1993. LNCS, vol. 697, pp. 438–449. Springer, Heidelberg (1993)

9. Holzmann, G., Peled, D., Yannakakis, M.: On nested depth first search. In: The Spin Verification System, pp. 23–32. American Mathematical Society, Providence (1996)

10. Holzmann, G.J.: The SPIN Model Checker. Pearson Education, London (2003)

11. Holzmann, G.J., Peled, D.: An improvement in formal verification. In: FORTE, pp. 197–211 (1994)

12. Johnson, C.G.: Genetic programming with fitness based on model checking. In: Ebner, M., O'Neill, M., Ekárt, A., Vanneschi, L., Esparcia-Alcázar, A.I. (eds.) EuroGP 2007. LNCS, vol. 4445, pp. 114–124. Springer, Heidelberg (2007)

13. Katz, G., Peled, D.: Genetic programming and model checking: Synthesizing new mutual exclusion algorithms. In: Cha, S(S.), Choi, J.-Y., Kim, M., Lee, I., Viswanathan, M. (eds.) ATVA 2008. LNCS, vol. 5311, pp. 33–47. Springer, Heidelberg (2008)

14. Katz, G., Peled, D.: Model checking-based genetic programming with an application to mutual exclusion. In: Ramakrishnan, C.R., Rehof, J. (eds.) TACAS 2008. LNCS, vol. 4963, pp. 141–156. Springer, Heidelberg (2008)

15. Katz, G., Peled, D.: Model checking driven heuristic search for correct programs. In: Peled, D.A., Wooldridge, M.J. (eds.) MoChArt 2008. LNCS, vol. 5348, pp. 122–131. Springer, Heidelberg (2009)

16. Katz, S., Peled, D.: Defining conditional independence using collapses. Theor. Comput. Sci. 101(2), 337–359 (1992)

17. Kinnear Jr., K.E.: Evolving a sort: Lessons in genetic programming. In: IJCNN, vol. 2, pp. 881–888 (1993)

18. Koza, J.R.: Genetic Programming: On the Programming of Computers by Means of Natural Selection. MIT Press, Cambridge (1992)

19. Kurshan, R.P., McMillan, K.L.: A structural induction theorem for processes. In: PODC, pp. 239–247 (1989)

20. Lamport, L.: A theorem on atomicity in distributed algorithms. Distributed Computing 4, 59–68 (1990)

21. Le Lann, G.: Distributed systems - towards a formal approach. In: IFIP Congress, pp. 155–160 (1977)

22. Lichtenstein, O., Pnueli, A.: Checking that finite state concurrent programs satisfy their linear specification. In: POPL, pp. 97–107 (1985)

23. Niebert, P., Peled, D., Pnueli, A.: Discriminative model checking. In: Gupta, A., Malik, S. (eds.) CAV 2008. LNCS, vol. 5123, pp. 504–516. Springer, Heidelberg (2008)

24. Oppen, D.C.: A $2^{2^{2^{pn}}}$ upper bound on the complexity of presburger arithmetic. J. Comput. Syst. Sci. 16(3), 323–332 (1978)

25. Overman, W.T., Crocker, S.D.: Verification of concurrent systems: Function and timing. In: PSTV, pp. 401–409 (1982)

26. Peterson, G.L.: An O(n log n) unidirectional algorithm for the circular extrema problem. ACM Trans. Program. Lang. Syst. 4(4), 758–762 (1982)

27. Pnueli, A.: The temporal logic of programs. In: FOCS, pp. 46–57 (1977)

28. Pnueli, A., Rosner, R.: On the synthesis of a reactive module. In: POPL, pp. 179–190 (1989)

29. Sistla, A.P., Clarke, E.M.: The complexity of propositional linear temporal logics. J. ACM 32(3), 733–749 (1985)

30. Suzuki, I.: Proving properties of a ring of finite state systems. Inf. Process. Lett. 28(4), 213–314 (1988)
31. Vardi, M.Y., Wolper, P.: Automata theoretic techniques for modal logics of programs. In: STOC, pp. 446–456 (1984)
32. Wolper, P., Lovinfosse, V.: Verifying properties of large sets of processes with network invariants. In: Sifakis, J. (ed.) CAV 1989. LNCS, vol. 407, pp. 68–80. Springer, Heidelberg (1990)

Stacking-Based Context-Sensitive Points-to Analysis for Java

Xin Li and Mizuhito Ogawa

School of Information Science,
Japan Advanced Institute of Science and Technology, Nomi, Japan

Abstract. Points-to analysis for Java infers heap objects that a reference variable can point to. Existing practiced context-sensitive points-to analyses are cloning-based, with an inherent limit to handle recursive procedure calls and being hard to scale under deep cloning. This paper presents a *stacking-based* context-sensitive points-to analysis for Java, by deriving the analysis as weighted pushdown model checking problems. To generate a tractable model for model checking, instead of passing global variables as parameters along procedure calls and returns, we model the heap memory with a global data structure that stores and loads global references with synchronized points-to information on-demand. To accelerate the analysis, we propose a two-staged iterative procedure that combines *local exploration* for lightening most of iterations and *global update* for guaranteeing soundness. In particular, *summary transition rules* that carry cached data flows are carefully introduced to trigger each local exploration, which boosts the convergence with retaining the precision. Empirical studies show that, our analysis scales well to Java benchmarks of significant size, and achieved in average $2.5X$ speedup in the two-staged analysis framework.

1 Introduction

The notion of context-sensitivity bears a similarity to inline expansion, as if method calls are replaced with bodies of the callees. As such, the typical *cloning-based* program analysis [17] creates a separate copy of a method call within a bounded call depth or with collapsing recursive procedure calls. The cloning-based approach has an inherit limit to handle (recursive) procedure calls. An alternative to obtaining context-sensitivity in terms of *valid call paths* is to model the program's call stack with the pushdown stack. Since the stack can grow unboundedly, no restriction is placed on the call depth and recursions. By valid, it means a procedure call always returns to the most recent call site.

Points-to analysis (PTA) infers the set of heap objects that a reference variable may point to. PTA for Java is featured for being interdependent of call graph construction, due to dynamic language features like late binding. The long-standing challenge is to design a scalable yet precise PTA. Context-sensitivity is shown to be crucial to the precision of PTA for Java. To the best of our knowledge, existing practiced PTA for Java are all cloning-based [17,13]. However,

K. Namjoshi, A. Zeller, and A. Ziv (Eds.): HVC 2009, LNCS 6405, pp. 133–149, 2011.

empirical study recently shows that, more than one thousand of methods are typically contained within recursive procedure calls in practice [18], and approximating recursions potentially threatens the analysis precision [6].

This paper presents a *stacking-based* context-sensitive PTA for Java, by encoding the analysis as model checking problems on WPDSs [11]. Our analysis is context-sensitive, field-sensitive, and flow-insensitive, with the call graph constructed on-the-fly. In contrast to the cloning-based approach, there is a single copy for each procedure in the analysis, while calling contexts are entirely characterized as (regular) configurations over the pushdown stack. Our first step to scalability is that, instead of passing global variables explicitly as parameters along procedure calls and returns (that is hopeless to scale from our empirical study) [10], we model the heap memory with a global data structure during the analysis, which loads intermediate points-to information of global references, and stores cached values to global references on-demand when they are referred to inside procedures. This encoding dramatically reduces the number of pushdown transitions and generates a tractable model for model checking.

To further accelerate the analysis, we propose a two-staged iterative procedure, denoted by $(\text{LE} \circ \text{GU})^*$ as opposed to the traditional iterative procedure denoted by GU^*, which combines *local exploration* (LE) for lightening most of iterative cycles and *global update* (GU) for guaranteeing the completeness. Our insight is, to localize most of iterative cycles on the partial program models in LEs, and perform GUs on the entire program model as few times as possible. In particular, *summary transition rules* that carry previously computed data flows are introduced to effectively trigger each LE and boost the convergence. In effect, the computation of data flows to some program point in the partial program model is divided into independent phases via *frontiers*: the computation of data flows from the program entry to frontiers and the computation of data flows from frontiers to the concerned program point. By carefully adding summary transition rules to frontiers, the analysis by $(\text{LE} \circ \text{GU})^*$ retains the same precision as the analysis by GU^*. Empirical studies show that, a substantial speedup in practice can be achieved by the two-staged analysis.

This paper primarily makes the following contributions.

- We present a scalable stacking-based context-sensitive PTA for Java by model checking WPDSs, with no restriction on (recursive) procedure calls.
- We propose a two-staged iteration procedure, supported by carefully introducing summary transition rules, to effectively accelerate the analysis.
- We implemented the analysis algorithms as a tool named *Japot*. Empirical study shows that *Japot* scales well to Java benchmarks of significant size.

The rest of the paper is organized as follows. Section 2 briefly reviews weighted pushdown model checking. Section 3 presents Java semantics and abstractions. Detection of points-to information by model checking is in Section 4. Section 5 presents a two-staged iteration procedure. Section 6 gives experiments and Section 7 discusses related work. Section 8 concludes the paper.

2   Weighted Pushdown Model Checking

Definition 1. *Define a **pushdown system** $P = (Q, \Gamma, \Delta, q_0, \omega_0)$, where Q is a finite set of states called control locations, Γ is a finite stack alphabet, and $\Delta \subseteq Q \times \Gamma \times Q \times \Gamma^*$ is a finite set of transition rules. $q_0 \in Q$ and $\omega_0 \in \Gamma^*$ are the initial control location and stack contents respectively. A transition rule $(p, \gamma, q, \omega) \in \Delta$ is denoted by $\langle p, \gamma \rangle \hookrightarrow \langle q, \omega \rangle$. A **configuration** of P is a pair $\langle q, \omega \rangle$ for $q \in Q$ and $\omega \in \Gamma^*$. Δ defines a transition relation $\Rightarrow$ on configurations such that $\langle p, \gamma\omega' \rangle \Rightarrow \langle q, \omega\omega' \rangle$ for each $\omega' \in \Gamma^*$ if $\langle p, \gamma \rangle \hookrightarrow \langle q, \omega \rangle$.*

Definition 2. *$S = (D, \oplus, \otimes, \mathbf{0}, \mathbf{1})$ with $\mathbf{0}, \mathbf{1} \in D$ is a **bounded idempotent semiring** if*

1. *$(D, \oplus)$ is a commutative monoid with $\mathbf{0}$ as its unit element, and $\oplus$ is idempotent, i.e., $a \oplus a = a$ for $a \in D$;*
2. *$(D, \otimes)$ is a monoid with $\mathbf{1}$ as the unit element;*
3. *$\otimes$ distributes over $\oplus$;*
4. *$\forall a \in D, a \otimes \mathbf{0} = \mathbf{0} \otimes a = \mathbf{0}$;*
5. *The partial ordering $\sqsubseteq$ is defined on D such that $\forall a, b \in D, a \sqsubseteq b$ iff $a \oplus b = a$.*

There are no infinite descending chains in D.

Definition 3. *Define a **weighted pushdown system** (WPDS) $W = (P, S, f)$, where $P = (Q, \Gamma, \Delta, q_0, \omega_0)$ is a pushdown system, $S = (D, \oplus, \otimes, \mathbf{0}, \mathbf{1})$ is a bounded idempotent semiring, and $f: \Delta \to D$ is a weight assignment function.*

When encoding the program as a WPDS, the bounded idempotent semiring models program data flows. A weight element encodes traditional program transformers; $f \oplus g$ combines data flows at the meet of control flows; $f \otimes g$ composes sequential control flows; $\mathbf{1}$ is identity function, and $\mathbf{0}$ implies program errors.

Definition 4. *Given a weighted pushdown system $W = (P, S, f)$, where $P = (Q, \Gamma, \Delta, q_0, \omega_0)$. Assume $\sigma = [r_0, ..., r_k]$ for $r_i \in \Delta (0 \le i \le k)$ to be a sequence of pushdown transition rules, and $v(\sigma) = f(r_0) \otimes ... \otimes f(r_k)$. Let **path**$(c, c')$ be the set of all transition sequences that transform configurations from c into c'. Given sets of regular configurations $C, C' \subseteq Q \times \Gamma^*$, for each configuration $c \in Q \times \Gamma^*$,*

- *the Generalized Pushdown Successor (**GPS**) problem is to find $gps(c) = \bigoplus\{v(\sigma) \mid \sigma \in \mathbf{path}(c', c), c' \in C\}$.*
- *the Generalized Pushdown Predecessor (**GPP**) problem is to find $gpp(c) = \bigoplus\{v(\sigma) \mid \sigma \in \mathbf{path}(c, c'), c' \in C'\}$.*
- *The Meet-Over-All-Valid-Path (**MOVP**) problem is to find $\text{MOVP}(C, C', W) = \bigoplus\{v(\sigma) \mid \sigma \in \mathbf{path}(c, c'), c \in C, c' \in C'\}$.*

Given $p \in Q, \gamma \in \Gamma$ and $c \in Q \times \Gamma^*$, and let $\mathbf{conf}(p, \gamma) = \{\langle p, \gamma\omega \rangle \mid \omega \in \Gamma^*\}$, further define $\widehat{\text{MOVP}}(c, \langle p, \gamma \rangle, W) = \text{MOVP}(\{c\}, \mathbf{conf}(p, \gamma), W)$.

Efficient algorithms for solving the GPS and GPP problems are proposed based on the fact that a regular set of configurations is closed under forward and backward reachability [11]. Then, MOVP is solved based on the results of either GPS or GPP. There are two off-the-shelf implementations of weighted pushdown model checking, Weighted PDS Library[1] and WPDS++ [3,4]. We exploit the former as the back-end analysis engine.

Example 1. As shown in Table 1, a context-sensitive PTA is able to distinguish that, x_1 and x_2 points to objects created at line 2 and 3, respectively. In contrast, an imprecise analysis may mix them.

Table 1. A Java Code Snippet

```
0.  public class Main {
1.     public static void main(String[] args){      8.   public static Object f₁(Object a){
2.        Object y₁ = new String();                  9.      return f₂(a);
3.        Object y₂ = new Object();                  10.   }
4.        Object x₁ = f₁(y₁);                        11.   public static Object f₂(Object b){
5.        Object x₂ = f₁(y₂);                        12.      return b;
6.        System.out.println(x₁.equals(x₂));         13.   }
7.     }                                             14. }
```

3   Semantics and Abstraction

3.1   Java Semantics on the Heap and Call Stack

Definition 5. *A method signature consists of method name, parameter types, and return type. We denote by $\mathcal{C}$ the set of classes, and denote by Ψ the set of method signatures. A method is identified by a pair of its enclosing class $C \in \mathcal{C}$ and its method signature $\psi \in \Psi$, denoted by $C.\psi$. The set of method identifiers is denoted by $\mathcal{C}.\Psi \subseteq \mathcal{C} \times \Psi$. We denote by Θ the class environment, including all classes, type representations of classes, and the type hierarchy.*

In Java, a heap object is a dynamically created instance of either a class or an array. Reference variables are typically local variables, method parameters, array references, and static or instance fields that hold reference types. Fields and array references can be regarded as global variables. A local variable v from its enclosing method $C.\psi$ is denoted by indexing with the scope as $v^{C.\psi}$. If $C.\psi$ is clear from the context, we often simply write v.

Definition 6. *The set of references is denote by $\mathcal{V}$, and the set of heap objects is denoted by $\mathcal{O}$. An abstract heap environment $\boldsymbol{henv}$ is a mapping, denoted by $\mapsto$, from $\mathcal{V}$ to $\mathcal{O}$. The set of abstract heap environments is denoted by Λ, on which the update operation $\odot$ is defined such that for $r, r' \in \mathcal{V}$, $o \in \mathcal{O}$, $(\boldsymbol{henv} \odot [r \mapsto o])r' = o$ if $r = r'$ and $(\boldsymbol{henv} \odot [r \mapsto o])r' = \boldsymbol{henv}(r')$ otherwise.*

---

[1] http://www.fmi.uni-stuttgart.de/szs/tools/wpds/

Definition 7. *We denote by $\mathcal{L}$ the set of program line numbers and denote by $\mathcal{S}$ the set of program statements. Let $\mathit{Stmt} : \mathcal{L} \to \mathcal{S}$ be the function that returns the statement at a given line number. $\mathcal{S}_\epsilon \subseteq \mathcal{S}$ denotes the set of statements that do not contain explicit method invocations and operate on the heap memory, and by $\mathcal{S}_I \subseteq \mathcal{S}$ denotes the set of statements that contains explicit method invocations.*

Definition 8. *Let $\mathit{Elem} = \mathcal{L} \times (\mathcal{O} \cup \{*\}) \times \mathcal{C}.\Psi$. Let $\Pi = \mathit{Elem}^*.\{\bot\}$ be the set of calling histories over the call stack. Define $\mathbf{push}(\mathit{stack}, e) = e.\mathit{stack}$ for $\mathit{stack} \in \Pi$; and $\mathbf{pop}(e.\mathit{stack}) = \mathit{stack}$, $\mathbf{top}(e.\mathit{stack}) = e$ for $e \in \mathit{Elem}$ and $\mathit{stack} \in \Pi$; and $\mathbf{pop}(\bot) = \mathbf{top}(\bot) = \bot$.*

A call stack symbol $\langle l, o, C.\psi \rangle \in \mathtt{Elem}$ denotes the program execution point at line l of the instance method $C.\psi$ that is invoked on the object o, and $\langle l, *, C.\psi \rangle \in \mathtt{Elem}$ represents an execution point inside a static method $C.\psi$.

Table 2. Transition Rules on the Heap and Call Stack

$\mathtt{stmt}(l)$ from $C.\psi$	henv'	stack'
$x = \mathtt{new}\ T$	$\mathtt{henv} \odot [x \mapsto \nu(\mathtt{henv})]$	
$x = y$	$\mathtt{henv} \odot [x \mapsto \mathtt{henv}(y)]$	$\mathbf{push}(s, e)$ where
$x := (\mathtt{T})y$	$\mathtt{henv} \odot [x \mapsto \mathtt{henv}(y)]$	$s = \mathbf{pop}(stack)$
$x := \mathtt{@this} :\ T$	$\mathtt{henv} \odot [x \mapsto \mathtt{henv}(\mathtt{this})]$	$e = \langle \mathtt{next}(l), o, C.\psi \rangle$
$x := \mathtt{@parameter}_k :\ T$	$\mathtt{henv} \odot [x \mapsto \mathtt{henv}(\mathtt{arg}_k)]$	$o =' *'$ if $C.\psi$ is static
$x = y[i]$	$\mathtt{henv} \odot [x \mapsto \mathtt{henv}(y)[i]]$	$o \in \mathtt{henv}(\mathtt{this}^{C.\psi})$ o.w.
$y[i] = x$	$\mathtt{henv} \odot [\mathtt{henv}(y)[i] \mapsto \mathtt{henv}(x)]$	
$x = y.f$	$\mathtt{henv} \odot [x \mapsto \mathtt{henv}(\mathtt{henv}(y).f)]$	
$y.f = x$	$\mathtt{henv} \odot [\mathtt{henv}(y).f \mapsto \mathtt{henv}(x)]$	
$\mathtt{return}\ y$	$\mathtt{henv} \odot [\mathtt{ret} \mapsto \mathtt{henv}(y)]$	$\mathbf{pop}(stack)$
$z = r_0.m(r_1, ..., r_n)$	$\mathtt{henv} \odot [\mathtt{this}^{C'.\psi'} \mapsto \mathtt{henv}(r_0)]$ $\odot [\mathtt{arg}_1^{C'.\psi'} \mapsto \mathtt{henv}(r_1)] \odot \cdots$ $\odot [\mathtt{arg}_n^{C'.\psi'} \mapsto \mathtt{henv}(r_n)]$ $\odot [z \mapsto \mathtt{ret}^{C'.\psi'}]$ where $\psi' \in \Psi$ is the method signature of m, and $C'.\psi' = \mathtt{resolve}(\mathtt{TypeOf}(o'), \psi', \Theta)$ for $o' = \mathtt{henv}(r_0)$.	$\mathbf{push}(s', e')$ where $s' = \mathbf{push}(s, e)$ $s = \mathbf{pop}(stack)$ $e = \langle \mathtt{next}(l), o, C.\psi \rangle$ $e' = \langle l_0^{C'.\psi'}, o', C'.\psi' \rangle$
$z = C'.m(r_1, ..., r_n)$	$\mathtt{honv} \odot [\mathtt{arg}_1^{C'.\psi'} \mapsto \mathtt{henv}(r_1)] \odot \cdots$ $\odot [\mathtt{arg}_1^{C'.\psi'} \mapsto \mathtt{henv}(r_1)]$ $\odot [z \mapsto \mathtt{ret}^{C'.\psi'}]$ where $\psi' \in \Psi$ is the method signature of m, and $C'.\psi' = \mathtt{resolve}(C', \psi', \Theta)$.	$\mathbf{push}(s, e')$ where $s = \mathbf{push}(\mathbf{pop}(stack), e)$ $e = \langle \mathtt{next}(l), o, C.\psi \rangle$ $e' = \langle l_0^{C'.\psi'}, *, C'.\psi' \rangle$

We use a transition system $(\mathtt{states}, s_{init}, \to)$ to represent the operational Java semantics on the heap and call stack, where $\mathtt{states} \subseteq \mathcal{L} \times \Theta \times \Lambda \times \Pi$ is the set of program states, each of which is a tuple of program locations, class environment, heap environments and calling histories, and $s_{init} \in \mathtt{states}$ is the initial state; $\to\, \subseteq \mathtt{states} \times \mathtt{states}$ is the set of transition rules.

As far as single-threaded Java programs are concerned, the next program location after each execution step at l ($\in \mathcal{L}$) is uniquely determined, and is denoted by $\text{next}(l)$. As given in Table 2, for the program execution of the statement $\text{stmt}(l)$ at $l \in \mathcal{L}$ from the method $C.\psi$, the transition rule is $\langle l, \Theta, \text{henv}, \text{stack} \rangle \to \langle \text{next}(l), \Theta, \text{henv}', \text{stack}' \rangle$. Here, ν is a function that generates a fresh heap object. this, arg_k and ret are fresh variables to denote the *this* reference of a class instance, the kth method argument, and the return variable, respectively. $\text{TypeOf} : \mathcal{O} \to \mathcal{C}$ is the function that returns the runtime type of a heap object. $\text{resolve} : \mathcal{C} \times \Psi \times \Theta \to \mathcal{C}.\Psi$ is the function implements how JVM resolves and the method to be invoked at runtime according to its method signature and possible enclosing class. $l_0^{C'.\psi'}$ refers to the entry point of the method $C'.\psi'$.

3.2 Abstraction

We apply the following abstractions to abstract away various sources of infinities.

- A unique abstract heap object models concrete heap objects created at the same allocation site (a.k.a., the context-insensitive heap abstraction). Thus, the number of abstract heap objects are syntactically bounded to be finite. An abstract heap object is a pair of its *allocation site* and *runtime type*.
- The indices of arrays are ignored, such that members of an array are not distinguished. We denote by $[\![o]\!]$ the unique representative for all members of the array instance o ($\in \mathcal{O}$). After abstracting the set of heap objects to be finite, the nesting of array and field reference become finite correspondingly.

Definition 9. *The set of abstract heap objects is $\text{Obj} = (\mathcal{L} \cup \{\_\}) \times \mathcal{C}$, where $\_$ is a fresh symbol for indicating nowhere. Let $\text{TypeOf} : \text{Obj} \to \mathcal{C}$ be the function returns the second projection of an abstract heap object.*

Definition 10. *Let $\text{RetPoint} \subseteq \mathcal{L} \times C.\Psi$ be the set of return points for method invocations. Define abstractions $\alpha_t : \text{Elem} \to \{\_\} \times C.\Psi$, $\alpha_r : \text{Elem} \to \text{RetPoint}$, and $\alpha_o : \text{Elem} \to \text{Obj} \times C.\Psi$ such that, for $\langle l, o, C.\psi \rangle \in \text{Elem}$,*

- $\alpha_t(\langle l, o, C.\psi \rangle) = (\_, C.\psi)$,
- $\alpha_r(\langle l, o, C.\psi \rangle) = (l, C.\psi)$, *and*
- $\alpha_o(\langle l, o, C.\psi \rangle) = (o, C.\psi)$.

α_r *and α_o are extended to the call stack in an element-wise manner.*

Definition 11. *Let $\mathbb{C} = (\{\_\} \times (C.\Psi)).\text{RetPoint}^*$ be the set of abstract calling contexts. Define a calling context abstraction $\alpha : \Pi \to \mathbb{C}$ such that $\alpha(\bot) = \epsilon$ and $\alpha(e.\text{stack}) = \alpha_t(e).\alpha_r(\text{stack})$ for $e \in \text{Elem}$, and $\text{stack} \in \Pi$.*

By Def. 10, α_t abstracts the topmost stack symbol for flow-insensitivity. α_r abstracts the return points of method invocations, which results in calling contexts in terms of call site strings. Our choice of α_r indicates, method invocations to the same method from different places of the same caller is still distinguished. An alternative of α_r is α_o, which abstracts calling contexts as sequences of heap objects on which methods are invoked, also known as object-sensitivity [9].

Definition 12. *We denote by* Ref *the set of abstract reference variables, and by* cc(v) *all possible abstract calling contexts for a reference variable* $v \in$ Ref.

Note that, cc(v) are exactly all possible calling contexts for the method which v belongs to, and cc(v) is automatically computed as the set of reachable regular configurations during model checking (Section 4).

Definition 13. *Let* $\mathbf{R} :$ Ref$\times \mathbb{C} \rightarrow \mathcal{P}(Obj)$ *be the function that stores the points-to relation, where* $\mathcal{P}$ *denotes the powerset operator.* $\mathbf{R} \downarrow_V : V \times \mathbb{C} \rightarrow \mathcal{P}(Obj)$ *is the restriction of* $\mathbf{R}$ *to* $V \subseteq$ Ref. *Define* $\ominus : \mathbf{R} \times \mathbf{R} \rightarrow \mathbf{R}$, *such that for any* $r \in$ Ref *and* $cc \in \mathbb{C}$, $(\mathbf{R}_1 \ominus \mathbf{R}_2)(r, cc) = \mathbf{R}_1(r, cc) \setminus \mathbf{R}_2(r, cc)$.

4   Detecting Points-to Information by Model Checking

This section presents, given points-to information and a call graph, how to detect new points-to information and enlarge a call graph in a context/field-sensitive and flow-insensitive way.

Definition 14. $\mathcal{G} = (M, \mathbf{E})$ *is a call graph of a program if* $M \subseteq \mathcal{C}.\Psi$ *and* $\mathbf{E} \subseteq M \times \mathcal{L} \times M$. *An element in* $\mathbf{E}$ *is called a call edge.*

We call $\mathbf{P} \subseteq \mathcal{C}.\Psi$ a program coverage, and denote the program coverage consisting of enclosing methods of program entries by $\mathbf{P}_0$.

Definition 15. *Let* Henv, sp *be fresh symbols. Define a weighted pointer assignment graph (WPAG)* $G = (N, L, \leadsto, n_0)$, *where* $N \subseteq ($Ref $\cup \{$Henv$\}) \times (\mathcal{C}.\Psi \cup$ RetPoint$)$ *is the set of nodes,* $L \subseteq \{\lambda x.\{o\} \mid o \in$ Obj$\} \cup \{\lambda x.x\}$ *is the set of labels,* $\leadsto \subseteq N \times L \times N$ *is the set of edges, and* $n_0 = ($Henv, sp$) \in N$ *is the root.*

A WPAG G is a directed labeled graph to represent data flow of heap objects. Henv and sp indicate the program environment that provides new abstract heap objects and program inputs, and the dummy program entry, respectively. The first projection of N represents abstract references, and the second projection represents their program scopes. Edges of G are classified into inter-edges ($\leadsto_i$, defined in Table 3) and intra-edges ($\leadsto_c, \leadsto_r, \leadsto_t$, defined in Table 4). An edge $(v, m) \leadsto (v', m')$ is denoted by

$$\begin{cases} \leadsto_i & \text{if } m = m' \in \mathcal{C}.\Psi \\ \leadsto_c & \text{if } m \neq m' \text{ and } m, m' \in \mathcal{C}.\Psi \end{cases} \qquad \begin{cases} \leadsto_r & \text{if } m' \in \text{RetPoint} \\ \leadsto_t & \text{if } m \in \text{RetPoint} \end{cases}$$

The procedure of finding new points-to information consists of the following steps. Step 1 builds intra-edges of a WPAG G given $\mathbf{R}$. Step 2 builds inter-edges of G given $\mathbf{E}$. During Step 2, the set of return points associated with each method invocation is recorded as a mapping $\mathbb{M} :\leadsto_c \rightarrow \mathcal{P}($RetPoint$)$. Initially, $\mathbb{M}$ is the constant function to the empty set. Step 3 encodes G as a WPDS W and detects new points-to information, denoted by $\widehat{\mathbf{R}}$, by model checking.

Table 3. $\mathcal{A}[\![\_]\!] : \mathcal{S}_\epsilon \to \mathcal{P}(\rightsquigarrow_i)$

$\mathcal{A}[\![x = \texttt{new T}]\!] = \{(\texttt{Henv}, C.\psi) \overset{\lambda x.\{(l,\texttt{T})\}}{\rightsquigarrow_i} (x, C.\psi)\}$

$\mathcal{A}[\![x = y]\!] = \{(y, C.\psi) \rightsquigarrow_i (x, C.\psi)\}$

$\mathcal{A}[\![x := (\texttt{T})y]\!] = \{(y, C.\psi) \rightsquigarrow_i (x, C.\psi)\}$

$\mathcal{A}[\![x := @\texttt{this} : \texttt{T}]\!] = \{(\texttt{this}, C.\psi) \rightsquigarrow_i (x, C.\psi)\} \cup A_e$

where $A_e = \{(\texttt{Henv}, C.\psi) \overset{\lambda x.\{(\_,\texttt{T})\}}{\rightsquigarrow_i} (\texttt{this}, C.\psi)\}$ if $C.\psi \in \mathbf{P}_0$ and $A_e = \emptyset$ otherwise

$\mathcal{A}[\![x := @\texttt{parameter}_k : \texttt{T}]\!] = \{(\texttt{arg}_k, C.\psi) \rightsquigarrow_i (x, C.\psi)\} \cup A_p$

where $A_p = \{(\texttt{Henv}, C.\psi) \overset{\lambda x.\{(\_,\texttt{T})\}}{\rightsquigarrow_i} (\texttt{arg}_k, C.\psi)\}$ if $C.\psi \in \mathbf{P}_0$ and $A_p = \emptyset$ otherwise

$\mathcal{A}[\![\texttt{return } x]\!] = \{(x, C.\psi) \rightsquigarrow_i (\texttt{ret}, C.\psi)\}$

$\mathcal{A}[\![x = y[i]]\!] = \{([\![o]\!], C.\psi) \rightsquigarrow_i (x, C.\psi) \mid o \in \mathbf{R}(y, \texttt{cc}(y))\} \cup A_g$

$\mathcal{A}[\![y[i] = x]\!] = \{(x, C.\psi) \rightsquigarrow_i ([\![o]\!], C.\psi) \mid o \in \mathbf{R}(y, \texttt{cc}(y))\} \cup A_g$

where $A_g = \{(\texttt{Henv}, C.\psi) \overset{\lambda x.s}{\rightsquigarrow_i} ([\![o]\!], C.\psi) \mid o \in \mathbf{R}(y, \texttt{cc}(y)), s = \mathbf{R}([\![o]\!], \texttt{cc}([\![o]\!]))\}$

$\mathcal{A}[\![x = y.f]\!] = \{(o.f, C.\psi) \rightsquigarrow_i (x, C.\psi) \mid o \in \mathbf{R}(y, \texttt{cc}(y))\} \cup A_f$

$\mathcal{A}[\![y.f = x]\!] = \{(x, C.\psi) \rightsquigarrow_i (o.f, C.\psi) \mid o \in \mathbf{R}(y, \texttt{cc}(y))\} \cup A_f$

where $A_f = \{(\texttt{Henv}, C.\psi) \overset{\lambda x.s}{\rightsquigarrow_i} (o.f, C.\psi) \mid o \in \mathbf{R}(y, \texttt{cc}(y)), s = \mathbf{R}(o.f, \texttt{cc}(o.f))\}$

Step 1: Building Intra-Procedural Data Flows

Table 3 gives rules that translate statements from $\mathcal{S}_\epsilon$ at line $l (\in \mathcal{L})$ of the method $C.\psi$ to intra-edges of G, denoted by $\mathcal{A}[\![\_]\!] : \mathcal{S}_\epsilon \to \mathcal{P}(\rightsquigarrow_i)$. For simplicity, we omit a weight associated to $\rightsquigarrow$ if it is $\lambda x.x$. Our modeling is featured as follows.

- In contrast to cloning-based approach, there is the unique abstract reference of each local reference variable. Global references are cloned only for methods inside which they are referred.
- Instead of passing global variables explicitly as parameters along procedure calls and returns, the heap memory is modelled with the global data structure $\mathbf{R}$ and provides global references with *cached data flows* (i.e., A_g, A_f) when they are locally referred (only necessary for field read).

Step 2: Building Inter-Procedural Data Flows

Table 4 gives rules that translate statements from $\mathcal{S}_I$ that contains explicit method invocations to inter-edges of G, denoted by $\mathcal{A}[\![\_]\!] : \mathcal{S}_I \to \rightsquigarrow_c \cup \rightsquigarrow_r \cup \rightsquigarrow_t$, where A_c denotes call edges, A_r denotes return edges, and A_t denotes data flows from return points to the calling procedure. Note that, $\texttt{Henv}$ as the program environment is explicitly passed as a parameter among calls and returns. During generation of inter edges, the mapping $\mathbb{M}$ is updated with newly produced return points. The translation rules for static method invocations can be defined similarly. Finally, new edges $\{n_0 \rightsquigarrow (\texttt{Henv}, C.\psi) \mid n_0 = (\texttt{Henv}, \texttt{sp}), C.\psi \in \mathbf{P}_0\}$ are added to G that lead from the dummy root node n_0 to the program entries.

Step 3: Building the WPDS W from G and Model Checking

Definition 16. *Let* $D_1 = \{\lambda x.s \mid s \in \mathcal{P}(Obj)\}$ *and* $D_2 = \{\lambda x.x \cup s \mid s \in \mathcal{P}(Obj)\}$. *Define a bounded idempotent semiring* $S = (D, \oplus, \otimes, \mathbf{0}, \mathbf{1})$, *such that*

Table 4. $\mathcal{A}[\![\_]\!] : \mathcal{S}_I \to \leadsto_c \cup \leadsto_r \cup \leadsto_t$

$$\mathcal{A}[\![z = r_0.f(r_1, ..., r_n)]\!] = A_c \cup A_r \cup A_t$$
where $A_c = \{(r_0, C.\psi) \leadsto_c (\mathtt{this}^{C'.\psi'}, C'.\psi')\} \cup \{(\mathtt{Henv}, C.\psi) \leadsto_c (\mathtt{Henv}, C'.\psi')\}$
$\qquad \cup \bigcup_{r_i \in \mathbf{Ref}} \{(r_i, C.\psi) \leadsto_c (\mathtt{arg}_i^{C'.\psi'}, C'.\psi')\}$
$\qquad A_r = \{(\mathtt{ret}^{C'.\psi'}, C'.\psi') \leadsto_r (\mathtt{ret}^{C'.\psi'}, \mathtt{rp})\} \cup \{(\mathtt{Henv}, C'.\psi') \leadsto_r (\mathtt{Henv}, \mathtt{rp})\}$
$\qquad A_t = \{(\mathtt{ret}^{C'.\psi'}, \mathtt{rp}) \leadsto_t (z, C.\psi)\} \cup \{(\mathtt{Henv}, \mathtt{rp}) \leadsto_t (\mathtt{Henv}, C.\psi)\}$
$\qquad \psi'$ is the method signature of the method f, and
$\qquad (C.\psi, l, C'.\psi') \in \mathbf{E}$, and $\mathtt{rp} = (l, C.\psi)$, and
$\qquad$ for all $r \in A_c$, $\mathbb{M}(r) = \mathbb{M}(r) \cup \{\mathtt{rp}\}$

- *The weighted domain $D = D_1 \cup D_2 \cup \{\boldsymbol{0}\}$, and $\boldsymbol{1} = \lambda x.x$;*
- $d_1 \otimes d_2 = d_1 \oplus d_2 = \lambda x.\ d_1(x) \cup d_2(x)$ *for* $d_1, d_2 \in D \setminus \{\boldsymbol{0}\}$
- $d \otimes \boldsymbol{0} = \boldsymbol{0} \otimes d = \boldsymbol{0}$ *for $d \in D$;*

It is easy to see that both the distributivity of $\otimes$ over $\oplus$ and the associativity of $\otimes$ hold. D_1 consists of constant functions, and $\lambda x.s \in D_1$ is that a reference points to the set of abstract heap objects s; and $\lambda x.x \cup s \in D_2$ is that a reference may keep unchanged along a path and be updated to point to s along another.

Given a WPAG G from Definition 15, a WPDS $W = (P, S, f)$ with $P = (Q, \Gamma, \Delta, q_0, w_0)$ is encoded G as follows,

- The set of control locations Q is the first projection of N, i.e., $\mathbf{Ref} \cup \{\mathtt{Henv}\}$;
- The stack alphabet Γ is the second projection of N, i.e., $C.\Psi \cup \mathtt{RetPoint}$;
- S is from Definition 16;
- $q_0 = \mathtt{Henv}$ and $w_0 = \mathtt{sp}$;
- For each edge r represented as $(v_1, m_1) \overset{l}{\leadsto} (v_2, m_2)$ such that $f(r) = l$ and
    - $\langle v_1, m_1 \rangle \hookrightarrow \langle v_2, m_2 \rangle$ if $r \in \leadsto_i$ or $\leadsto_t$;
    - $\langle v_1, m_1 \rangle \hookrightarrow \langle v_2, m_2 m_r \rangle$ for each $m_r \in \mathbb{M}(\leadsto_c)$ if $r \in \leadsto_c$;
    - $\langle v_1, m_1 \rangle \hookrightarrow \langle v_2, \epsilon \rangle$ if $r \in \leadsto_r$.

Definition 17. *Let $W = (P, S, f)$ be a weighted pushdown system with $P = (Q, \Gamma, \Delta, q_0, \gamma_0)$. For any reference $v \in \mathbf{Ref}$ from the method $C.\psi$, $\widehat{\boldsymbol{R}}(v, \mathtt{cc}(v)) = \widehat{MOVP}(c, \langle v, C.\psi \rangle, W)\ (H_{init}(v))$, where $c = \langle q_0, \gamma_0 \rangle$ and H_{init} denotes the initial abstract heap environment such that $H_{init}(v) = \emptyset$ for any $v \in \mathbf{Ref}$.*

To solve $\mathtt{MOVP}(C_s, C_t, W)$, we (i) first compute $\mathbf{gps}(c)$ for each $c \in C_t$ given C_s, and then (ii) read out and combine the value of all paths between C_s and C_t. We denote by $H = 2^{|\mathtt{Obj}|}$ the length of the longest descending chain of the weighted domain, and by T the time to perform either $\otimes$ or $\oplus$. In our case, the time required to perform step (ii) can be ignored, and the worst case time complexity of performing step (i) is $\mathcal{O}(|Q|^2\, |\Delta|\, |\Gamma|\, H\, T)$.

5   Acceleration by Lightening Iterative Cycles

5.1   A Traditional Iterative Procedure Scheme

Algo. `OnTheFlyPTA` in Fig. 1 sketches a procedure scheme for on-the-fly Java PTA. It starts with analyzing the program entry points $\mathbf{P}_0$, and computes the call graph $\mathbf{E}$ and points-to relation $\mathbf{R}$ until convergence. For each iterative cycle,

- `FindPointsTo` : $\mathbf{P} \times \mathbf{E} \times \mathbf{R} \to \mathbf{R}$ (line 3) detects points-to information $\widehat{\mathbf{R}}$ on the partial program $\mathbf{P}$, according to updated information in the previous iteration. The updated points-to information $\Delta\mathbf{R}$ is derived at line 4.
- `FindCallEdges` : $\mathbf{P} \times \mathbf{R} \times \Theta \to \mathbf{E}$ (line 5) resolves call relation $\widehat{\mathbf{E}}$ according to $\widehat{\mathbf{R}}$, obeying to the standard JVM semantics. The updated call relation $\Delta\mathbf{E}$ is derived at line 6.
- `TakeReachables` : $\mathbf{E} \to \mathbf{P}$ (line 8) returns the set of methods (reachable from program entries) to be analyzed in the next iteration. It can be defined as the union of the first and third projection of $\mathbf{E}$.

> **Algorithm** `OnTheFlyPTA`
> **Input**: the program entry points $\mathbf{P}_0$ and the class environment Θ
> **Output**: $\mathcal{G} = (M, \mathbf{E})$ and $\mathbf{R}$
> 0. $\mathbf{E} := \emptyset$; $\mathbf{R} := \emptyset$; $\mathbf{P} := \mathbf{P}_0$
> 1. **do**
> 2. $\widehat{\mathbf{R}} = \text{FindPointsTo}(\mathbf{P}, \mathbf{E}, \mathbf{R})$
> 3. $\Delta\mathbf{R} := \widehat{\mathbf{R}} \ominus \mathbf{R}$
> 4. $\mathbf{R} := \mathbf{R} \sqcup \Delta\mathbf{R}$
> 5. $\widehat{\mathbf{E}} := \text{FindCallEdges}(\mathbf{P}, \mathbf{R}, \Theta)$
> 6. $\Delta\mathbf{E} := \widehat{\mathbf{E}} \setminus \mathbf{E}$
> 7. $\mathbf{E} := \mathbf{E} \cup \Delta\mathbf{E}$
> 8. $\mathbf{P} := \text{TakeReachables}(\mathbf{E})$
> 9. **while** $\Delta\mathbf{E} \neq \emptyset$ or $\Delta\mathbf{R} \neq \emptyset$

Fig. 1. A Procedure for On-the-fly Java Points-to Analysis

Theorem 1. *The algorithm* `OnTheFlyPTA` *terminates if (i) the domain of* $\boldsymbol{P}$, $\mathbf{E}$ *and* $\mathbf{R}$ *are finite, and (ii) each of these functions* `FindPointsTo`, `FindCallEdges` *and* `TakeReachables` *is monotonic wrt the set inclusion on* $\boldsymbol{P}$, $\mathbf{E}$ *and the element-wise extension on* $\mathbf{R}$ *of the set inclusion.*

`FindPointsTo` is the core procedure of PTA for Java. For most cloning-based algorithms, `FindPointsTo` corresponds to the propagation of points-to sets, which is typically reduced to constraint solving problems. In contrast, we derive the analysis algorithm as model checking problems on WPDSs (Section 4). Since the abstraction given in Section 3.2 is an over approximation, soundness of our analysis is straightforward.

5.2   A Two-Staged Iterative Procedure Scheme

For an on-the-fly points-to analysis, the program coverage is enlarged when points-to analysis proceeds. However, we found that only part of the whole program would effectively contribute to the enlargement of the program coverage. To boost on-the-fly PTA, we propose a two-staged iterative procedure, denoted by $(\text{LE} \circ \text{GU})^*$, which combines two phases of LE (*local exploration*) and GU (*global update*). Generally, an LE iteration localizes the analysis on small parts of the program, which is more likely to enlarge the program coverage, and GU is performed on-demand for guaranteeing completeness. Line 9 switches LE and GU when conditions defined by `SwitchCond` are satisfied. Otherwise, a GU iteration will be triggered to check sound convergence.

> **Algorithm** `TwoStaged_OnTheFlyPTA`
> **Input**: the program entry points $\mathbf{P}_0$ and the class environment Θ
> **Output**: $\mathcal{G} = (\mathcal{M}, \mathbf{E})$ and $\mathbf{R}$
> 0. $\mathbf{E} := \emptyset;\ \mathbf{R} := \emptyset;\ \mathbf{P} := \mathbf{P}_0;\ \text{NotDone} = \text{NotDone'} := true$
> 1.   **do**
> 2. $\text{NotDone} := \text{NotDone'}$
> 3. $\widehat{\mathbf{R}} = \texttt{FindPointsTo}(\mathbf{P}, \mathbf{E}, \mathbf{R})$
> 4. $\Delta\mathbf{R} := \widehat{\mathbf{R}} \ominus \mathbf{R}$
> 5. $\mathbf{R} := \mathbf{R} \sqcup \Delta\mathbf{R}$
> 6. $\widehat{\mathbf{E}} := \texttt{FindCallEdges}(\mathbf{P}, \mathbf{R}, \Theta)$
> 7. $\Delta\mathbf{E} := \widehat{\mathbf{E}} \setminus \mathbf{E}$
> 8. $\mathbf{E} := \mathbf{E} \cup \Delta\mathbf{E}$
> 9.   **if** $\texttt{SwitchCond}(\Delta\mathbf{E}, \Delta\mathbf{R}) = true$ **then**
> 10. $\mathbf{P} := \texttt{TakeCoverage}(\mathbf{E}, \Delta\mathbf{E}, \Delta\mathbf{R})$
> 11. $\text{NotDone'} := true$
> 12.   **else**
> 13. $\mathbf{P} := \texttt{TakeReachables}(\mathbf{E})$
> 14. $\text{NotDone'} := false$
> 15.   **while** $\Delta\mathbf{E} \neq \emptyset$ or $\Delta\mathbf{R} \neq \emptyset$ or $\text{NotDone} = true$

Fig. 2. A Two-Staged Procedure for On-the-fly Java Points-to Analysis

Definition 18. $\texttt{SwitchCond}(\Delta\boldsymbol{E}, \Delta\boldsymbol{R}) = (\Delta\boldsymbol{E} \neq \emptyset) \vee (\Delta\boldsymbol{R} \downarrow_{Ref_f} \neq \emptyset)$, *where* $Ref_f \subseteq Ref$ *is the set of base references of instance fields.*

Def. 18 means that, an LE is triggered when either new call edges are detected or new global references are found. Both indicates that the underlying model for model checking is extended with new pushdown transitions.

Definition 19. $\texttt{TakeCoverage}(\boldsymbol{E}, \Delta\boldsymbol{E}, \Delta\boldsymbol{R}) = M_1 \cup M_2 \cup M_3 \cup \texttt{TakeReachables}(\Delta\boldsymbol{E})$,

$M_1 = \left\{ m'' \ \middle| \ \begin{array}{l} \exists m, m' \in \mathcal{C}.\Psi \ \exists l, l' \in \mathcal{L}.\ (m, l, m') \in \Delta\boldsymbol{E}, (m, l', m'') \in \boldsymbol{E}, \\ \text{and the return type of } m'' \text{ is a reference type} \end{array} \right\}$

$M_2 = \{ C.\psi \mid v^{C.\psi} \in Ref_f \text{ and } \Delta\boldsymbol{R}(v^{C.\psi}, \texttt{cc}(v^{C.\psi})) \neq \emptyset \}$

$M_3 = \{ m, C.\psi \mid \Delta\boldsymbol{R}(\texttt{ret}^{C.\psi}, \texttt{cc}(\texttt{ret}^{C.\psi})) \neq \emptyset \text{ and } \exists l \in \mathcal{L}.(m, l, C.\psi) \notin \boldsymbol{E} \}$

A partial model is taken in the ways defined in `TakeCoverage`, where M_1 says that, if a new call relation found from m to m', other callees of m that returns values of reference type are collected. M_2 says that, if the points-to information of base variables of instance fields are updated, their enclosing methods are collected. M_3 says that, if the points-to information of return variables of $C.\psi$ is updated, $C.\psi$ and methods that call $C.\psi$ and are not included in the previous LE are collected. Note that, our choice of `TakeCoverage` is inspired and decided by empirical studies on practiced Java benchmarks regarding efficiency.

Theorem 2. *The algorithm* `TwoStaged_OnTheFlyPTA` *terminates if (i) the domain of* $\mathbf{P}$, $\mathbf{E}$ *and* $\mathbf{R}$ *are finite, and (ii) each of these functions* `FindPointsTo`, `FindCallEdges`, `TakeCoverage` *and* `TakeReachables` *is monotonic on all arguments from their domains, and (iii)* `SwitchCond`$(\emptyset, \emptyset) =$ false.

5.3   Adding Summary Transition Rules in LE

As given in Table 5 that extends translation rules Table 5, to make the two-staged iterative procedure work effectively, *summary transition rules* (i.e., A_c) that carry cached data flows to $\mathbf{P}_E$ (Def. 20) are introduced, when building a WPAG in an LE iteration. Translation rules leading from the dummy root node n_0 to the program entries are lifted to $\{n_0 \rightsquigarrow (\texttt{Henv}, C.\psi) \mid n_0 = (\texttt{Henv}, \texttt{sp}), C.\psi \in \mathbf{P}_E\}$.

Definition 20. *Given a program coverage* $\mathbf{P}$ *and the call relation* $\mathbf{E}$, $\mathbf{P}_E = \{m \in \mathbf{P} \mid m' \notin \mathbf{P} \text{ if } (m', m) \in \mathbf{E}\}$.

Table 5. $\mathcal{B}[\![\_]\!] : \mathcal{S}_\epsilon \to \mathcal{P}(\rightsquigarrow_i)$

$\mathcal{B}[\![x := @\texttt{this} : \ \mathbf{T}]\!] = \mathcal{A}[\![x := @\texttt{this} : \ \mathbf{T}]\!] \cup A_c$
where $A_c = \{(\texttt{Henv}, C.\psi) \overset{\lambda x.s}{\rightsquigarrow}_i (\texttt{this}, C.\psi) \mid s = \mathbf{R}(\texttt{this}, \texttt{cc}(\texttt{this}))\}$
if $C.\psi \in \mathbf{P}_E$ and $A_c = \emptyset$ otherwise
$\mathcal{B}[\![x := @\texttt{parameter}_k : \ \mathbf{T}]\!] = \mathcal{A}[\![x := @\texttt{parameter}_k : \ \mathbf{T}]\!] \cup A_c$
where $A_c = \{(\texttt{Henv}, C.\psi) \overset{\lambda x.s}{\rightsquigarrow}_i (\texttt{arg}_k, C.\psi) \mid s = \mathbf{R}(\texttt{arg}_k, \texttt{cc}(\texttt{this}))\}$
if $C.\psi \in \mathbf{P}_E$ and $A_c = \emptyset$ otherwise

In sequel, we show that adding summary transitions to arguments (i.e., $\texttt{arg}_i$, `this`) of methods from $\mathbf{P}_E$ will not cause any loss of precision.

Definition 21. *For a pushdown system* $P = (Q, \Gamma, \Delta, q_0, \omega_0)$, *define* $\unrhd \subseteq Q \times \Gamma \times Q \times \Gamma$, *such that* $\langle p, \gamma \rangle \unrhd \langle p', \gamma' \rangle$ *if there exists* $\langle p, \gamma \rangle \Rightarrow^* \langle p', \gamma' \omega' \rangle$ *for some* $\omega' \in \Gamma^*$. *Define* $\unrhd_s \subseteq \unrhd$ *such that* $\langle p, \gamma \rangle \unrhd_s \langle p', \gamma' \rangle$ *if (i)* $\langle p, \gamma \rangle \unrhd \langle p', \gamma' \rangle$ *and (ii) for each* $\omega \in \Gamma^*$ *and all transition sequence of* $\sigma : \langle p, \gamma \omega \rangle \Rightarrow^* \langle p', \gamma' \omega' \omega \rangle$ *for some* $\omega' \in \Gamma^*$, *and any* $\langle p'', \omega'' \rangle$ *appearing in* σ *satisfies* $|w''| > |w|$. $\langle p, \gamma \rangle$ *is a* **dominator** *of* $\langle p', \gamma' \rangle$ *if* $\langle p, \gamma \rangle \unrhd_s \langle p', \gamma' \rangle$.

Definition 22. *Let $W = (P, S, f)$ be a WPDS with $P = (Q, \Gamma, \Delta, q_0, \gamma_0)$. For $p \in Q, \gamma \in \Gamma$, $\{\langle p_i, \gamma_i \rangle \mid 0 \leq i \leq k\}$ is a **dominator set** of $\langle p, \gamma \rangle$ if (1), for each i with $0 \leq i \leq k$, $\langle p_i, \gamma_i \rangle \unrhd_s \langle p, \gamma \rangle$, and (2), for each transition sequence $\sigma : \langle q_0, \gamma_0 \rangle \Rightarrow^* \langle p, \gamma w \rangle$ with $w \in \Gamma^*$, there uniquely exists $\langle p_j, \gamma_j \rangle$ such that $\langle p_j, \gamma_j w' \rangle$ for some $w' \in \Gamma^*$ appears in σ.*

Lemma 1. *Given a WPDS $W = (P, S, f)$ where $P = (Q, \Gamma, \Delta, q_0, \gamma_0)$. For $p \in Q, \gamma \in \Gamma$, let $\mathbb{H}$ be a dominator set of $\langle p, \gamma \rangle$ and let $c = \langle q_0, \gamma_0 \rangle$, we have*
$$\widehat{MOVP}(c, \langle p, \gamma \rangle, W) = \bigoplus\nolimits_{\langle p_i, \gamma_i \rangle \in \mathbb{H}} \widehat{MOVP}(c, \langle p_i, \gamma_i \rangle, W) \otimes \widehat{MOVP}(\langle p_i, \gamma_i \rangle, \langle p, \gamma \rangle, W).$$

Proof. Straightforward by definitions of frontiers and MOVP problems.

By Lemma 1, the computation of MOVP problems can be soundly divided into two independent phases via dominators.

Definition 23. *Given a WPDS $W = (P, S, f)$ with $P = (Q, \Gamma, \Delta, q_0, \gamma_0)$. For $p \in Q \setminus \{q_0\}, \gamma \in \Gamma \setminus \{\gamma_0\}$, $\langle p, \gamma \rangle$ is a **frontier** of W if either $\langle p, \gamma \rangle \unrhd_s \langle p', \gamma' \rangle$ or $\langle p, \gamma \rangle \not\unrhd \langle p', \gamma' \rangle$ for any $p' \in Q, \gamma' \in \Gamma$. A **frontier set** of W, denoted by $\mathbb{F}_W$, is a set of frontiers and $\langle p, \gamma \rangle \in \mathbb{F}_W$ implies $\langle p', \gamma' \rangle \notin \mathbb{F}_W$ if $\langle p, \gamma \rangle \unrhd_s \langle p', \gamma' \rangle$.*

Theorem 3. *Given a WPDS $W = (P, S, f)$ with $P = (Q, \Gamma, \Delta, q_0, \gamma_0)$. Let $W' = (P', S, f)$ with $P' = (Q', \Gamma', \Delta' \cup \delta, q_0, \gamma_0)$, where $Q' \subseteq Q$, $\Gamma' \subseteq \Gamma$, $\Delta' \subseteq \Delta$, and $\delta = \{r = \langle q_0, \gamma_0 \rangle \hookrightarrow \langle p_i, \gamma_i \rangle, f'(r) = \widehat{MOVP}(c, \langle p_i, \gamma_i \rangle, W) \oplus f(r) \mid \langle p_i, \gamma_i \rangle \in \mathbb{F}_{W'}\}$. For $p \in Q', \gamma \in \Gamma'$, let $c = \langle q_0, \gamma_0 \rangle$, $\widehat{MOVP}(c, \langle p, \gamma \rangle, W') \sqsupseteq \widehat{MOVP}(c, \langle p, \gamma \rangle, W)$.*

Proof. By Definition 23, given $p \in Q, \gamma \in \Gamma$, any frontier set $\mathbb{F}_{W'}$ of W' can be decomposed into disjoint union $\mathbb{F}_{W'} = F_1 \uplus F_2$, where F_1 is some collection of dominators of $\langle p, \gamma \rangle$ and $F_2 \subseteq \{\langle p', \gamma' \rangle \mid \widehat{MOVP}(\langle p', \gamma' \rangle, \langle p, \gamma \rangle, W') = \mathbf{0}\}$. The proof is done according to Lemma 1 and the fact that F_1 may not contain a dominator set of $\langle p, \gamma \rangle$.

By Theorem 3, the analysis of LEs with introducing summary transition rules will never cause any loss of precision, but can be not complete. The completeness is guaranteed by analysis of GUs.

Note that, the set of arguments $C_E = \{(\mathtt{arg}_k, C.\psi), (\mathtt{this}, C.\psi) \mid C.\psi \in \mathbf{P}_E\}$ from $\mathbf{P}_E$ is a witness of the frontier set $\mathbb{F}_{W'}$, where W' is the WPDS encoded from methods of $\mathbf{P}$ augmented with cached transition rules. Recall the example in Table 1, assume the partial model $\mathbf{P}$ taken in an LE consists of methods f_1 and f_2. We know $\mathbf{P}_E$ consists of f_1 only by definition. A loss of precision would be incurred, if summary transition rules A_c are introduced to arguments in f_2.

6 Empirical Studies

We developed our analysis algorithms as a tool named Japot[2], which exploits Soot2.3.0 [16] for preprocessing from Java programs to Jimple codes, and the

---

[2] As an approximation, return variables from any native methods and reflection calls can point to objects whose type allows, and a throw exception can be handled by any exception handler whose declared type allows as an over-approximation.

Weighted PDS Library as the model checking engine. We perform experiments on Java applications from the Ashes benchmark suite [15] and the influential DaCapo benchmark suite [2] (Table 6). These applications are de facto benchmarks when investigating Java points-to analysis. We target on the newest version of DaCapo benchmark which requires JDK 1.5 or above, and stable Ashes benchmarks for which JDK 1.3 suffices. In sequel, the performance of Japot is measured by *call graph generation* in terms of the number of reachable methods, which is given in the "# Reachable Methods" column and these numbers take into account libraries used by each benchmark. Benchmarks on which the back-end model checker runs out of memory are not shown. All experiments were performed on a Mac OS X v.10.5.2 with a Xeon 2×2.66 GHz Dual-Core processor, and 4GB RAM. Only one processor is used in the following experiments.

Table 6. Benchmark Statistics and Call Graph Generation

Benchmark	# Reachable Methods			# Statements	# Suite	# JDK
	CHA	Japot	Prec.↑	Japot		
soot-c	5460	5079	7%	83936		
sablecc-j	13,055	9004	31%	143140	Ashes	JDK 1.3.1_01
antlr	10,728	×				
bloat	12,928	11090	14%	194063		
chart	30,831	×				
jython	14,603	×				
pmd	12,485	×			Dacapo	JDK 1.5.0_13
hsqldb	9983	8394	16%	142629		
xalan	9977	8392	16%	141405		
luindex	10,596	8961	15%	152592		
lusearch	11,190	9580	14%	163958		
eclipse	12,703	×				

Table 7. An Acceleration on Efficiency

Benchmark	# (LE ∘ GU)* (sec.)	# GU* (sec.)	# Acceleration
soot-c	656	1591	59%
sablecc-j	1547	2785	44%
bloat	12339	41434	70%
hsqldb	1205	2910	59%
xalan	1321	2926	55%
luindex	1514	3880	61%
lusearch	1757	4057	57%

The sub-column titled CHA gives the number of reachable methods by the CHA (Class Hierarchy Analysis) of Spark in soot-2.3.0. The sub-column titled Japot gives results computed by our context-sensitive PTA and the "# Statements" column gives the number of Jimple statements that Japot analyzed. The

"Prec.↑" sub-column shows how much improvement on precision can be obtained by Japot over CHA. Our proposals regarding program modelling alone does not yield high scalability. Applying *type filtering* on-the-fly as usual and ignoring differences of string constants are also essential to the scalability.

We studied the efficiency improvement of $(LE \circ GU)^*$ over GU^*, and initial results are given in Table 7[3]. The "# $(LE \circ GU)^*$" column and the "# GU^*" column gives the time in seconds of performing these iterative schemes respectively. The "# Acceleration" column shows an acceleration in terms of $\frac{|GU^*| - |(LE \circ GU)^*|}{|GU^*|}$, which shows that $(LE \circ GU)^*$ is $2.5X$ faster in average than GU^*. We expect novel strategies of taking LEs can improve the practical efficiency even more.

7   Related Work

One of the pioneer work is Andersen's PTA for C [1]. It is a subset-based, flow-insensitive analysis encoded as constraint solving problems, such that object allocations and pointer assignments are described by subset constraints, e.g. $x = y$ induces $pta(y) \subseteq pta(x)$. The scalability of Andersen's analysis has been greatly improved by more efficient constraint solvers. Andersen's analysis was introduced to Java by using annotated constraints [12].

The first scalable cloning-based context-sensitive Java PTA is presented in [17], in which programs and analysis problems are encoded as rules in logic program Datalog. Calling contexts are cloned after merging loops as equivalent classes. The BDD (Binary Decision Diagram) based implementation, as well as approximation by collapsing recursions, enable the analysis to scale. As discussed in [6], there are usually rich and large loops within the call graph, and the loss of precision is potentially incurred after approximating recursions.

Reps, et al. present a general framework for program analysis based on CFL-reachability [10], in which a PTA for C is shown by formulating pointer assignments as productions of context-free grammars. Inspired by this work, Sridharan, et al. formulated Andersen's analysis for Java [14] as balanced-parentheses problems regarding field read and write. A novel refinement-based analysis [13] is based on context-insensitive analysis and recovers the precision on-demand by removing imprecise propagation of points-to sets as violating a grammar for balanced parentheses, regarding both heap access and method calls. It shows good precision and scalability with respect to downcast safety analysis.

SPARK[5] is a widely-used test-bed for experimenting with Java PTA. It supports both equality and subset-based analysis, provides various algorithms for call graph construction, such as CHA, RTA(Rapid Type Analysis), and on-the-fly algorithms, as well as variations on field-sensitivity. The BDD-based implementation of the subset-based algorithms further improves the efficiency.

---

[3] Note that,data structures and program states cannot be shared between soot (in Java) and the back-end model checker (in C). These numbers include DISK IO time for exchanging information between these two parts via files.

One stream of research examines calling contexts in terms of sequences of objects on which methods are invoked, called *object-sensitivity* [9]. Similar to call-site strings based approach, the sequence of receiver objects can be unbounded and demands proper approximations, like k-CFA. [6] indicates that object-sensitivity excels at precision and is more likely to scale. Last but not the least, existing practiced Java PTA as discussed above, are cloning-based for context-sensitivity and have restrictions on handling recursive procedure calls.

In contrast to points-to analysis with call graph constructed on-the-fly, an ahead-of-time points-to analysis is proposed as one run of weighted pushdown model checking [7]. The notion of valid paths are enriched with further obeying to the Java semantics on dynamic dispatch. In particular, invalid control flows that violate Java semantics on dynamic dispatch are detected as those carrying conflicted data flows. The analysis enjoys context-sensitivities regarding both call graph construction and valid paths.

Last but not least, WPDSs are extended to conditional weighted pushdown systems (CWPDSs), by further associating each transition rule with a regular language that specifies conditions under which the transition rule can be applied [8]. There are wider applications of CWPDs when analyzing programs with objected-oriented features, for which WPDSs are not precise enough under a direct application. It is also shown that, the model checking problem on CWPDSs can be reduced to model checking problems on WPDSs.

8   Conclusions

We presented a scalable stacking-based context-sensitive points-to analysis for Java. The algorithm is derived as model checking problems on WPDSs, and no restriction is placed on (recursive) procedure calls. A two-staged iterative procedure is further proposed to effectively accelerate the analysis, supported by introducing summary transition rules. Our new iteration schemes shows the potential of an incremental points-to analysis, and we are extending the current setting with performing local explorations only.

Acknowledgments

This research is supported by STARC (Semiconductor Technology Academic Research Center). We thank Dr. Stefan for help us using the Weighted PDS Library. We also thank anonymous reviewers for their valuable comments.

References

1. Andersen, L.: Program analysis and specialization for the c programming language. PhD thesis (1994)
2. Blackburn, S.M., Garner, R., Hoffman, C., et al.: The DaCapo benchmarks: Java benchmarking development and analysis. In: Proceedings of the 21st Annual ACM SIGPLAN Conference on Object-Oriented Programing, Systems, Languages, and Applications, OOPSLA 2006, New York, NY, USA, pp. 169–190 (October 2006)

3. Lal, A., Reps, T.W.: Improving pushdown system model checking. In: Ball, T., Jones, R.B. (eds.) CAV 2006. LNCS, vol. 4144, pp. 343–357. Springer, Heidelberg (2006)
4. Lal, A., Reps, T.W.: Solving multiple dataflow queries using WPDSs. In: Alpuente, M., Vidal, G. (eds.) SAS 2008. LNCS, vol. 5079, pp. 93–109. Springer, Heidelberg (2008)
5. Lhoták, O., Hendren, L.: Scaling Java points-to analysis using Spark. In: Hedin, G. (ed.) CC 2003. LNCS, vol. 2622, pp. 153–169. Springer, Heidelberg (2003)
6. Lhoták, O., Hendren, L.: Context-sensitive points-to analysis: is it worth it? In: Mycroft, A., Zeller, A. (eds.) CC 2006. LNCS, vol. 3923, pp. 47–64. Springer, Heidelberg (2006)
7. Li, X., Ogawa, M.: An ahead-of-time yet context-sensitive points-to analysis for Java. In: Proceedings of BYTECODE 2009, York, ENTCS17798. Elsevier, Amsterdam (March 2009)
8. Li, X., Ogawa, M.: Conditional weighted pushdown systems and applications. In: Proceedings of the 2010 ACM SIGPLAN Workshop on Partial Evaluation and Program Manipulation (PEPM 2010), pp. 141–150. ACM, New York (2010)
9. Milanova, A., Rountev, A., Ryder, B.G.: Parameterized object sensitivity for points-to analysis for java. ACM Trans. Softw. Eng. Methodol. 14(1), 1–41 (2005)
10. Reps, T.: Program analysis via graph reachability. In: Proceedings of the 1997 International Symposium on Logic Programming, ILPS 1997, pp. 5–19. MIT Press, Cambridge (1997)
11. Reps, T., Schwoon, S., Jha, S., Melski, D.: Weighted pushdown systems and their application to interprocedural dataflow analysis. Sci. Comput. Program. 58(1-2), 206–263 (2005)
12. Rountev, A., Milanova, A., Ryder, B.G.: Points-to analysis for Java using annotated constraints. SIGPLAN Not. 36(11), 43–55 (2001)
13. Sridharan, M., Bodík, R.: Refinement-based context-sensitive points-to analysis for Java, vol. 41, pp. 387–400. ACM, New York (2006)
14. Sridharan, M., Gopan, D., Shan, L., Bodík, R.: Demand-driven points-to analysis for java. SIGPLAN Not. 40(10), 59–76 (2005)
15. Vallée-Rai, R.: Ashes suite collection, http://www.sable.mcgill.ca/ashes
16. Vallée-Rai, R., Gagnon, E., Hendren, L.J., Lam, P., Pominville, P., Sundaresan, V.: Optimizing Java bytecode using the Soot framework: Is it feasible? In: Watt, D.A. (ed.) CC 2000. LNCS, vol. 1781, pp. 18–34. Springer, Heidelberg (2000)
17. Whaley, J., Lam, M.S.: Cloning-based context-sensitive pointer alias analysis using binary decision diagrams. In: ACM SIGPLAN Conference on Programming Language Design and Implementation (PLDI 2004), pp. 131–144 (2004)
18. Xu, G., Rountev, A.: Merging equivalent contexts for scalable heap-cloning-based context-sensitive points-to analysis. In: Proceedings of the 2008 International Symposium on Software Testing and Analysis, ISSTA 2008, pp. 225–236. ACM, New York (2008)

An Interpolating Decision Procedure for Transitive Relations with Uninterpreted Functions[*]

Daniel Kroening[1] and Georg Weissenbacher (白傑岳)[1,2,**]

[1] Computing Laboratory, Oxford University
[2] Computer Systems Institute, ETH Zurich

Abstract. We present a proof-generating decision procedure for the quantifier-free fragment of first-order logic with the relations $=$, $\neq$, $\geq$, and $>$ and argue that this logic, augmented with a set of theory-specific rewriting rules, is adequate for bit-level accurate verification. We describe our decision procedure from an algorithmic point of view and explain how it is possible to efficiently generate Craig interpolants for this logic.

Furthermore, we discuss the relevance of the logical fragment in software model checking and provide a preliminary evaluation of its applicability using an interpolation-based program analyser.

1 Introduction

Interpolants play an ever more important role in software and hardware verification [1]. Since interpolants are typically constructed from proofs of inconsistency, interpolation-based verification techniques depend on efficient, proof-generating decision procedures. Interpolating decision procedures have been available for over a decade [2,3], but the field is still advancing rapidly.

McMillan's landmark paper [3] gives an axiomatic description of his interpolating theorem prover FOCI. Recently, more algorithmic descriptions of similar interpolating decision procedures have been published [4,5], indicating that publications that make the material in [3] more accessible are well appreciated.

We present a graph-based interpolating decision procedure for a subset of quantifier free first-order logic with a fixed set of relations, an extension of the logic covered by [4]. We support equality ($=$), disequality ($\neq$), and strong and weak inequality ($>$ and $\geq$, respectively). Furthermore, we provide limited support for interpreted functions such as bit-vector operations. Our presentation emphasises the algorithmic point of view.

Our work is motivated by the discrepancy between the bit-vector interpretation underlying most programming languages and the domains $\mathbb{R}$ or $\mathbb{Z}$ used by many interpolating decision procedures. The decision procedure we present

[*] Supported by the Semiconductor Research Corporation (SRC) under contract no. 2006-TJ-1539 and by the EU FP7 STREP MOGENTES (project ID ICT-216679).
[**] Supported by Microsoft Research's European PhD Scholarship Programme.

K. Namjoshi, A. Zeller, and A. Ziv (Eds.): HVC 2009, LNCS 6405, pp. 150–168, 2011.
© Springer-Verlag Berlin Heidelberg 2011

is *sound* for bit-level formulæ, i.e., if a formula is satisfiable in the bit-vector interpretation, then our algorithm will not conclude that it is unsatisfiable.

Our contribution is a self-contained, algorithmic description of a bit-level accurate decision procedure integrating rewriting rules for theory-specific axioms. We provide a preliminary evaluation of the suitability of our first-order logic fragment for software verification using a re-implementation of the model checking algorithm in [6].

2 Preliminaries

In the following, we define $\mathcal{L}$, a quantifier-free, conjunctive fragment of first-order logic. We restrict the predicates of this language to equality ($=$), disequality ($\neq$), and strong and weak inequality ($>$ and $\geq$, respectively).

Syntax. We fix an enumerable set of variables, function symbols, and constant symbols. Well-formed elements of $\mathcal{L}$ are generated by the following set of rules:

- A *term* t is a constant, a variable, or an application $f(t_1, \ldots t_n)$ of an n-ary function symbol f to terms $t_1, \ldots, t_n$.
- An *atom* $t_1 \triangleright t_2$ is a binary relation $\triangleright \in \{=, \geq, >, \neq\}$ applied to two terms t_1 and t_2. We do not allow any predicates other than the relations listed above.
- A *formula* F is a conjunction of atoms.

Note that the set of atoms is closed under negation, i.e., the negation $\neg(t_1 \triangleright t_2)$ of an atom can be expressed in terms of an atom. Conjunction ($\wedge$) is the only logical connective we allow in $\mathcal{L}$. This is a common restriction for (interpolating) decision procedures for specialised theories, since arbitrary propositional connectives can be handled using the orthogonal approach presented in [3,7].

Interpretations. We use the standard interpretation of the relation symbols $=$ and $\neq$. The relation $\geq$ is a partial order over the (interpreted) domain $\mathcal{D}$, and $t_i > t_j$ denotes $(t_i \geq t_j) \wedge (t_i \neq t_j)$. An *interpreted* n-ary function symbol f has a well-defined function $f^{\mathcal{M}} : \mathcal{D}^n \to \mathcal{D}$ associated to it, whereas an *uninterpreted* function symbol has no other property than its name and arity.

We use $F \models G$ to state that the formula F entails G.

Craig interpolation. Since $\mathcal{L}$ is a fragment of first-order logic, there exists a *Craig interpolant* (a first-order logic formula) for every inconsistent pair of $\mathcal{L}$-formulæ F and G:

Definition 1 (Craig interpolant for $\mathcal{L}$). *Given an unsatisfiable $\mathcal{L}$-formula $F \wedge G$, a Craig interpolant is a first-order logic formula I such that*

1. *$F \models I$,*
2. *$G \models \neg I$, and*
3. *the variables and function symbols I refers to are common to F and G.*

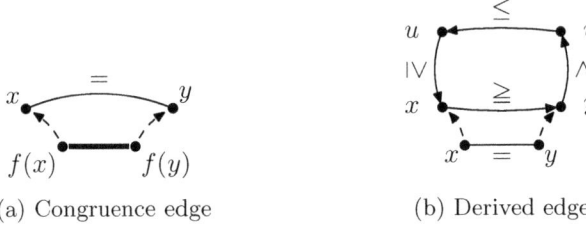

(a) Congruence edge   (b) Derived edge

Fig. 1. Congruence edges and derived edges

Remark 1. Note that I is not necessarily a $\mathcal{L}$-formula (and may not even be expressible in $\mathcal{L}$). An example is the pair of formulæ $f(x_0) \neq f(x_5) \wedge x_0 = x_1 \wedge x_2 = x_3 \wedge x_4 = x_5$ and $x_1 = x_2 \wedge x_3 = x_4$ and their interpolant $x_1 \neq x_2 \vee x_3 \neq x_4$.

Graph representation of $\mathcal{L}$ formulæ. The fact that an $\mathcal{L}$-formula F is a conjunction of atoms of the form $t_i \rhd t_j$ enables us to represent F using a graph [8].

Definition 2 ($\mathcal{L}$-graph). *Given a formula F, let $\mathcal{G}_F(V, E)$ be a directed graph, where each term t_i in F corresponds to a node v_i in V, and each atom $t_i \rhd t_j$ corresponds to a $\rhd$-labelled edge $(v_i \overset{\rhd}{\to} v_j) \in E$, $\rhd \in \{=, \geq, >, \neq\}$. Atoms $t_i \rhd t_j$ with a symmetric relation $\rhd \in \{=, \neq\}$ additionally contribute an edge $(v_j \overset{\rhd}{\to} v_i)$. For convenience, we use undirected edges to depict equalities and disequalities. In accordance to [3], we write $v_i \simeq v_j$ if and only if $i = j$.*

Due to the presence of functions in $\mathcal{L}$-terms, the congruence relation may give rise to additional equality edges in the graph: The congruence relation satisfies, in addition to the properties of the equality relation, the monotonicity axioms, i.e., for all n-ary functions f, it holds that $f(s_1, \ldots, s_n) = f(t_1, \ldots, t_n)$ whenever $s_i = t_i$ holds for all i in $\{1, .., n\}$. We use *congruence edges* to depict such equalities (see Fig. 1a). The dashed arrows indicate that $f(x) = f(y)$ is derived from the equality of the sub-terms $x = y$.

Definition 3 (Contradictory and equality-entailing cycles). *A contradictory cycle [8] in an $\mathcal{L}$-graph is a cyclic path consisting of either*

a) edges labelled with $=$ and a single edge labelled with $\neq$, or
b) edges labelled with either $=$ or $\geq$ and at least one edge labelled with $>$.

An equality-entailing cycle in an $\mathcal{L}$-graph is a cyclic path consisting of edges labelled with either $=$ or $\geq$. For any two terms t_i and t_j corresponding to nodes in an equality-entailing cycle, it holds that $t_i \geq t_j$ and $t_j \geq t_i$, and thus $t_i = t_j$.

We depict derived edges using a graphical representation similar to congruence edges (see Fig. 1b). In this example, the equality $x = y$ is derived from the equality-entailing cycle $x \overset{\geq}{\to} y \overset{\geq}{\to} v \overset{\geq}{\to} u \overset{\geq}{\to} x$.

3   A Graph-Based Decision Procedure

We begin this section with a brief outline of our decision procedure for $\mathcal{L}$-formulæ followed by a detailed description of the proof-generating algorithm. Let $\mathcal{G}(V, E)$ be the $\mathcal{L}$-graph for a given formula F. The decision procedure is subdivided into two phases:

1. In the first phase, the algorithm searches for contradictory or equality-entailing cycles with edges labelled $=$, $\geq$, and $>$ (Def. 3a) in the graph $\mathcal{G}(V, E^{\geq})$, where $E^{\geq}$ denotes $E \setminus \{(v_i \overset{\neq}{\to} v_j) \in E\}$. If a contradictory cycle exists, the algorithm terminates. Otherwise, the procedure adds to E the edges $v_i \overset{=}{\to} v_j$ and $v_j \overset{=}{\to} v_i$ for all nodes v_i, v_j adjacent in an equality-entailing cycle.
2. In the second phase, additional equalities are inferred by means of constant propagation and congruence closure and searches for contradictory cycles with edges labelled $=$ or $>$ (Def. 3b) in the graph $\mathcal{G}(V, E^{\neq})$, where $E^{\neq} = \{(v_i \triangleright v_j) \in E \,|\, \triangleright \in \{=, \neq\}\}$.

The phases are iterated until no new equalities can be inferred. Both phases use well-known and efficient graph algorithms such as Tarjan's algorithm for the computation of *strongly connected components* (SCCs) and a graph-based *union-find* data structure. In a pre-processing step, we form two (possibly non-disjoint) sets of the atoms in F, one of which contains the inequalities and equalities, and one which contains equalities and disequalities.

Phase I: Inequalities. Let $\mathcal{G}(V, E^{\geq})$ be the $\mathcal{L}$-graph corresponding to the equality and inequality atoms of F. Using Tarjan's algorithm, we compute all strongly connected components in $\mathcal{G}(V, E^{\geq})$ and classify them as contradictory or equality-entailing cycles, respectively:

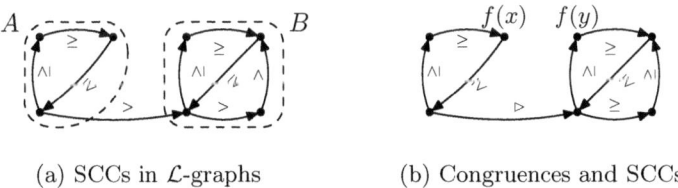

(a) SCCs in $\mathcal{L}$-graphs   (b) Congruences and SCCs

Fig. 2. Strongly connected components (SCCs) in $\mathcal{L}$-graphs

1. A SCC is contradictory if it contains at least one edge $v_i \overset{>}{\to} v_j$ (see component B in Fig. 2a). Then, any path from v_j to v_i forms a contradictory cycle with $v_i \overset{>}{\to} v_j$. If our algorithm finds a contradictory SCC, we compute the shortest such path and report it as a proof of inconsistency.

2. A SCC is equality-entailing if it contains no edge labelled with $>$ (see component A in Fig. 2a or the SCCs in Fig. 2b). In this case, we conclude that for any edge $v_i \overset{\triangleright}{\to} v_j$ in the SCC $t_i = t_j$ holds for the corresponding terms. The derived equalities are passed on to the second phase.

Phase II: Equalities and Disequalities. The second phase starts with computing the equivalence closure of the equality atoms (and the equalities derived in the first phase). For this purpose, we use a *proof-generating* union-find data structure that incrementally constructs an $\mathcal{L}$-graph $\mathcal{G}(V, E^=)$, where $E^=$ denotes a set of edges labelled with $=$. In the following, we present the modifications necessary to generate a proof of inconsistency. In a union-find data structure, each equivalence class corresponds to a sub-graph of $\mathcal{G}(V, E^=)$ identified by its *representative*, and each node which is not a representative holds a reference to its parent node (indicated by an directed edge in our illustrations). The data structure supports two operations:

1. $Find(v_i)$ returns the representative of the node v_i.
2. $Union(v_i, v_j)$ adds an (undirected) equality edge to the graph $\mathcal{G}(V, E^=)$ and merges the two equivalence classes containing v_i and v_j, respectively.

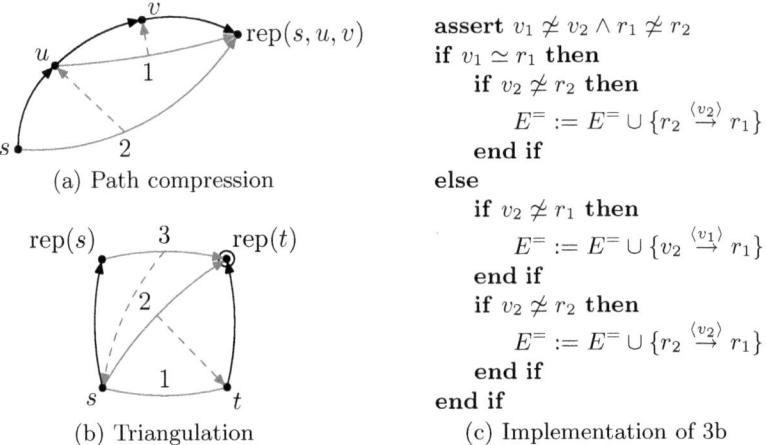

(a) Path compression

(b) Triangulation

```
assert v₁ ≉ v₂ ∧ r₁ ≉ r₂
if v₁ ≃ r₁ then
    if v₂ ≉ r₂ then
        E= := E= ∪ {r₂ →⟨v₂⟩ r₁}
    end if
else
    if v₂ ≉ r₁ then
        E= := E= ∪ {v₂ →⟨v₁⟩ r₁}
    end if
    if v₂ ≉ r₂ then
        E= := E= ∪ {r₂ →⟨v₂⟩ r₁}
    end if
end if
```

(c) Implementation of 3b

Fig. 3. An illustration of union-find operations

The $Find(v_i)$ operation performs *path compression* in order to reduce the computational effort in case of repeated queries for v_i. During this process, it adds new derived edges to $E^=$, which connect v_i directly with its representative. This is illustrated by the example in Fig. 3a. $Find$ follows the parent nodes until it reaches the representative. In Fig. 3a, the call to $Find(v_1)$ results in two recursive calls $Find(v_2)$ and $Find(v_3)$. The latter call returns v_4 as the representative for v_3. We add $v_2 \overset{\langle v_3 \rangle}{\to} v_4$ to $E^=$ (step 1 in Fig. 3a) and replace the parent v_3 with v_4. Here, the label $\langle v_3 \rangle$ is used to memorise the fact that $v_2 \overset{=}{\to} v_3$

derives from $v_2 \stackrel{=}{\to} v_3 \stackrel{=}{\to} v_4$ (visualised by the dashed arrow). Finally, $Find(v_2)$ yields v_4 and we add $v_1 \stackrel{\langle v_3 \rangle}{\to} v_4$ to $E^=$ and replace the parent v_2 with v_4. Thus, $Find(v_1)$ returns v_4.

The $Union(v_1, v_2)$ operation merges two equivalence classes with the representatives r_1, r_2 (obtained using $Find$). We assume that redundant unions are ignored, i.e., $v_1 \not\simeq v_2$ and $r_1 \not\simeq r_2$. Consider the example in Fig. 3b. We add the edge $v_1 \stackrel{=}{\to} v_2$ (step 1) and conclude that the terms corresponding to r_1 are r_2 equivalent. The algorithm chooses a new representative (r_1 in our example), favouring nodes with a higher in-degree. The resulting edge $v_2 \stackrel{\langle v_1 \rangle}{\to} r_1$ is labelled accordingly in step 2, in order to memorise its derivation. Finally, we connect r_2 and r_1; the corresponding edge derives from $r_2 \stackrel{=}{\to} v_2 \stackrel{=}{\to} r_1$.

Observe that $Union$ triangulates the sub-graph spanning $V = \{v_1, v_2, r_1, r_2\}$. Fig. 3c shows the general algorithm for this triangulation (where r_1 is the representative node with the higher in-degree), which is a constant time operation.

Using $Union$, we compute the equivalence closure for F by adding all equivalence atoms and derived equalities to $\mathcal{G}(V, E^=)$. We can now efficiently query whether a disequality $t_i \neq t_j$ contradicts the equality relations stored in $\mathcal{G}(V, E^=)$ by checking whether $Find(v_i) \simeq Find(v_j)$. If this is the case, we obtain a contradictory cycle $v_i \stackrel{\neq}{\to} v_j \stackrel{=}{\to} r \stackrel{=}{\to} v_i$. From this cycle, we obtain a proof for the inconsistency by repeatedly expanding derived edges $v_i \stackrel{\langle v_j \rangle}{\to} v_k$ to $v_i \stackrel{=}{\to} v_j \stackrel{=}{\to} v_k$. Edges derived in Phase I are justified by their respective equality-entailing cycles.

Congruence closure. The decision procedure described above lacks a provision for deriving congruence edges (Fig. 1a) and is therefore not sufficient to support uninterpreted functions. An equality relation $t_i = t_j$ in the congruence graph $\mathcal{G}(V, E^=)$ gives rise to a congruence edge representing $f(t_i) = f(t_j)$, which, in return, may entail additional equality relations in $\mathcal{G}(V, E^=)$. Therefore, we use an incremental *congruence closure* algorithm (following the ideas presented in [9]) that is closely intertwined with the construction of the $\mathcal{L}$-graph for equalities.

The algorithm uses the union-find data-structure representing $\mathcal{G}(V, E^=)$. It *indexes* each representative in $\mathcal{G}(V, E^=)$ with a term t_c. Thus, all terms in an equivalence class of $\mathcal{G}(V, E^=)$ are associated with the same term t_c. If the equivalence class contains an interpreted constant (e.g., a numeral), we choose it as the index term,[1] otherwise, we use the term corresponding to the representative of the equivalence class as index. In this setting, an equivalence class containing terms with function symbols represents a set of congruence relations. Consider two terms $f(t_i)$ and $f(t_j)$, where t_i and t_j have the same representative indexed with t_c. Then $f(t_i)$ and $f(t_j)$ belong to the same equivalence class.

In addition, we maintain a function $Lookup(f(t_c))$ which maps $f(t_c)$ to a term $f(t_i)$ such that t_i belongs to an equivalence class indexed with t_c, or $\bot$ if there is no such term in $\mathcal{G}(V, E^=)$.

---

[1] Note this constant is unique, since an equivalence class that contains two constants with a different interpretation contains a contradictory cycle.

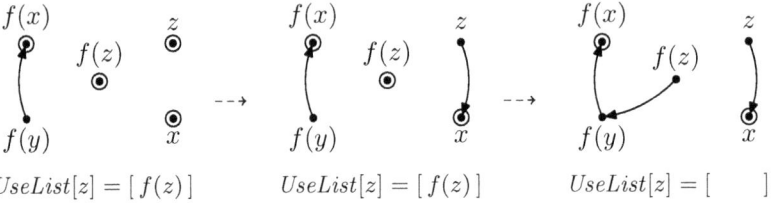

Fig. 4. A 3-step example illustrating the congruence closure algorithm

The *Union* operation potentially changes the representatives of the equivalence classes in $\mathcal{G}(V, E^=)$. Therefore, the algorithm maintains for each index term t_c a list *UseList(t_c)* of terms that contain a sub-term indexed with t_c. This list is updated whenever *Union* merges two equivalence classes. W.l.o.g., assume that *Union* merges an equivalence class indexed with t_c with an equivalence class indexed with t_c', choosing the latter term as the new index. Then, for each $f(t_i) \in UseList(t_c)$, where t_c is the index term associated with t_i, the algorithm proceeds as follows:

- If *Lookup*($f(t_c')$) returns $f(t_j)$, it uses *Union* to add a the congruence edge for $f(t_i) = f(t_j)$ to $\mathcal{G}(V, E^=)$ and memorises that the edge is derived from $t_i = t_j$. Furthermore, $f(t_i)$ is moved from *UseList(t_c)* to *UseList(t_c')*.
- If *Lookup*($f(t_c')$) returns $\perp$, it sets *Lookup*($f(t_c')$) to $f(t_i)$ and moves $f(t_i)$ to *UseList(t_c')*.

Example 1. Consider a union-find data structure with four equivalence classes $\{f(x), f(y)\}$, $\{f(z)\}$, $\{x\}$, and $\{z\}$ (see Fig. 4, on the left). *UseList*$[z]$ contains $f(z)$, since z is a sub-term of $f(z)$. Adding $x = z$ yields a new equivalence class $\{x\} \cup \{z\}$. Assume that the representative of the resulting equivalence class $\{x, z\}$ is x and that *Lookup*($f(x)$) = $f(y)$. Then the algorithm infers $f(z) = f(y)$.   ◁

The extension to n-ary functions is straight-forward. An efficient implementation based on *currifying* is presented in [9].

Bit-vector theory axioms, constant propagation, and interpreted functions. Our decision procedure provides limited support for the theory of bit-vectors by integrating a small set of bit-vector axioms and rewriting rules. Furthermore, whenever possible, it uses interpreted functions and constants in order to simplify terms. This is achieved by the following mechanisms:

1. We order all interpreted constants $c_1, \ldots, c_n$ processed in Phase I and add $n - 1$ inequality relations of the form $c_i < c_{i+1}$, $1 \leq i < n$ to $\mathcal{G}(V, E^{\geq})$ *before* computing the SCCs.
2. In Phase II, if *Union* is applied to two terms indexed with different interpreted constants c_1 and c_2, we introduce the disequality $c_1 \neq c_2$.
3. Let T be the set of terms corresponding to the nodes in $\mathcal{G}(V, E^=)$. For each $f(t_i) \in T$ such that f is an interpreted function symbol in a given theory $\mathcal{T}$

$$(t_2 + c) \neq t_2 \text{ if } c \neq 0 \bmod 2^m \qquad (t \ll c) = (t + t) \text{ if } c = 1$$
$$(t_2 + c) = t_2 \text{ if } c = 0 \bmod 2^m \qquad (t \ll c) = (2^c \cdot t) \text{ if } 1 < c < m$$

Fig. 5. Two examples for rewriting rules for m-bit variables

$$\frac{t_1 = t_2 \,\&\, t_3}{t_1 \leq t_2 \quad t_1 \leq t_3} \qquad \frac{t_1 = t_2 \,|\, t_3}{t_1 \geq t_2 \quad t_1 \geq t_3} \qquad \frac{t_1 + t_2 = t_1}{t_2 = 0}$$

Fig. 6. Examples of axioms for bit-vector operations

and t_i is a term indexed with an interpreted constant c, we check whether $f(c)$ can be simplified to a term t_j not containing any variables or function symbols that do not occur in $f(c)$. If this is the case, and $t_j \in T$ or t_j is an interpreted constant, we add the equivalence relations $f(t_i) = f(c)$ (derived from $t_i = c$) and $f(c) = t_j$ (a tautology in T) to $\mathcal{G}(V, E^=)$. This technique allows us to perform *bit-level-accurate* simplifications of terms.

4. We apply a fixed set of rewriting rules of the form $t \rhd t'$ to all terms t, where t' is the term obtained by applying the rule to t. All rules have the property that they do not introduce variables. Examples of such rules are listed in Fig. 5. If t and t' correspond to nodes in $\mathcal{G}(V, E^=)$, we add the relation $t \rhd t'$.

5. Axioms of the form $(t_1 \rhd_1 t_2) \vdash (t_3 \rhd_2 t_4)$ may be applied if t_3 and t_4 refer to a subset of the non-logical symbols in t_1 and t_2. Examples of such axioms are provided in Fig. 6.[2]

Combining both phases. As explained above, equality relations derived from equality-entailing cycles in Phase I are passed on to Phase II. Now consider the $\mathcal{L}$-graph in Fig. 2b. Adding the congruence edge corresponding to $f(x) = f(y)$ results in a new SCC, which, depending on the label $\rhd$ in Fig. 2b, is either contradictory or equality-entailing. Therefore, the congruence edges generated in Phase II must be added to $\mathcal{G}(V, E^{\doteq})$, necessitating an additional iteration of Phase I. The two phases need to be iterated until no more new congruence edges are generated. Since both phases are exchanging equalities exclusively, our implementation is essentially a Nelson-Oppen-style decision procedure.

Complexity. Tarjan's algorithm applied in Phase I has a run-time linear in the number n of edges of the graph. The computation of the equivalence closure in the second phase takes $O(n \cdot \alpha(n))$ time, where α is the inverse of the Ackermann function $A(n, n)$. The congruence closure is of complexity $O(n \cdot \log n)$ [9]. Thus, a single iteration of Phase I and Phase II takes $O(n \cdot \log n)$ time.

It remains to determine how often the phases need to be iterated. Since the algorithm never adds redundant congruence edges, the congruence closure adds at most $O(n)$ equalities (see [9]). Due to the restrictions on the application of

---

[2] The naïve application of such axioms increases the complexity of the algorithm significantly. Therefore, we apply each axiom only once in an initial rewriting phase.

rules and axioms, rewriting interpreted functions increases the number of sub-terms by at most a constant factor. Altogether, we face a run-time complexity of $O(n^2 \cdot \log n)$ for our decision procedure.

Finally, the extraction of an explanation from a contradictory cycle can be performed in $O(n \cdot \log n)$ time, since the derived edges form a tree.

Proofs of inconsistency. We review the artefacts generated by our decision procedure. A proof of inconsistency of an $\mathcal{L}$-formula F is a *contradictory cycle* comprising

- edges directly corresponding to relations in F,
- edges derived from equality-entailing cycles, and
- congruence edges, derived from a number of equality relations.

In the next section we explain how a Craig interpolant can be constructed from such a proof of inconsistency.

4   Extracting Interpolants from Contradictory Cycles

This section introduces the concept of coloured $\mathcal{L}$-graphs and explains how interpolants can be constructed from contradictory cycles in such a coloured graph.

Colouring $\mathcal{L}$-graphs. Given an $\mathcal{L}$-formula $F \wedge G$, we say that a node v_i of the corresponding graph $\mathcal{G}(V, E)$ is F-colourable if the corresponding term t_i refers only to variables and function symbols in F; similarly for G. We use V_F and V_G to refer to the set of F-colourable and G-colourable nodes, respectively. This definition splits $V = V_F \cup V_G$ into two non-disjoint sets of vertices. It leaves us a choice for a subset $V_S \stackrel{def}{=} (V_F \cap V_G)$ of V. We refer to V_S as *shared vertices*.

An edge $v_i \stackrel{\triangleright}{\to} v_j$ is F-colourable if and only if $\{v_i, v_j\} \subseteq V_F$; analogously for G. We use E_F (E_G) to refer to the F-colourable (G-colourable, respectively) edges in E. An edge is colourable if it is either F-colourable or G-colourable. The edges of the initial $\mathcal{L}$-graph $\mathcal{G}(V, E)$, in which each edge corresponds to an atom in $F \wedge G$, are always colourable. This is not necessarily the case for the graph that we obtain by computing the congruence closure (in Phase II). Consider the nodes labelled $f(x)$ and $f(y)$ in the $\mathcal{L}$-graph in Fig. 2b. Assume that the variable x occurs only in F and y occurs only in G. If we deduce $f(x) = f(y)$ from $x = y$, then the corresponding edge is not colourable.

It is, however, possible to transform a congruence-closed $\mathcal{L}$-graph into a colourable graph [4,10]. We provide a constructive proof based on structural induction over an $\mathcal{L}$-graph with congruence edges:

1. *Base case.* Colour the equality edges of the $\mathcal{L}$-graph according to their respective atoms in the formula $F \wedge G$.
2. *Induction step.* The argument is split into two cases:
    (a) *Derived edges.* For each edge $v_i \stackrel{=}{\to} v_j$ derived from an equality-entailing cycle, there exists an edge $v_i \stackrel{\triangleright}{\to} v_j$ ($\triangleright \in \{\geq, =\}$) in that cycle, which is, by the induction hypothesis, colourable. Let $v_i \stackrel{=}{\to} v_j$ take the colour of that edge.

Table 1. Rules for labelling contracted edges

	$=$	$\neq$	$\geq$	$>$
$=$	$=$	$\neq$	$\geq$	$>$
$\geq$	$\geq$	$\perp$	$\geq$	$>$
$>$	$>$	$\perp$	$>$	$>$

In order to label facts, the labels of the edges on a path are merged according to the rules to the left. By construction, the decision procedure described in Section 3 guarantees that no fact in a proof of inconsistency has an undefined ($\perp$) label.

(b) *Congruence edges.* Pick any non-colourable congruence edge with nodes $v_{f(x)}$ and $v_{f(y)}$ labelled $f(x)$ and $f(y)$, respectively. By the induction hypothesis, all edges in the path $v_x \to \ldots \to v_y$ entailing $x = y$ can be coloured. Since v_x and v_y are of different colour, there is a path prefix $v_x \to \ldots \to v_z$ such that all nodes in the prefix are of the same colour and $v_z \in V_S$. Let z be the term that corresponds to v_z. Then, the term $f(z)$ refers only to non-logical symbols common to F and G. Introduce a new node $v_{f(z)}$ representing $f(z)$ and add an equality edge $v_{f(x)} \to v_{f(z)}$ justified by $v_x \to \ldots \to v_z$, and a new congruence edge $v_{f(z)} \to v_{f(y)}$ justified by $v_z \to \ldots \to v_y$. All these new elements are colourable.

This proof translates into an algorithm of complexity $O(n \cdot \log n)$. The transformation yields a graph representing a formula *equisatisfiable* with $F \wedge G$, i.e., the modified graph contains a contradictory cycle if and only if the original congruence-closed graph $\mathcal{G}(V, E)$ contains one.

It is straight-forward to extend this argument to the edges introduced by the term rewriting rules and axioms in Section 3. Consider, w.l.o.g., an F-coloured node v_i corresponding to a term t, and a node v_j corresponding to the rewritten term t'. Due to the restriction that the rewriting rule $t \rightsquigarrow t'$ must not introduce new non-logical symbols,[3] the edge $v_i \to v_j$ can be coloured with 'F'. A similar argument holds for axioms, which do not change the colour of the affected edge.

This line of reasoning leads to the following observation:

Lemma 1. *A proof of inconsistency, which is a sub-graph of the congruence-closed $\mathcal{L}$-graph $\mathcal{G}(V, E)$ obtained using the algorithm in Section 3, can be transformed into a colourable graph.*

Furthermore, given that an $\mathcal{L}$-graph $\mathcal{G}(V, E)$ represents a formula $F \wedge G$, which is a *conjunction* of atoms, the formula represented by a sub-graph is implied by $F \wedge G$. Thus, the proof of inconsistency is implied by the original formula $F \wedge G$.

Interpolants from coloured inconsistency proofs. Given a coloured proof of inconsistency, it is now possible to *factorise* this graph according to the colour of its edges. Accordingly, a *factor* of a path in this graph is a maximal sub-path consisting of edges of equal colour. If we contract a factor $v_1 \overset{\triangleright_1}{\to} \ldots \overset{\triangleright_{n-1}}{\to} v_n$, we obtain a *fact* $v_1 \overset{\triangleright}{\to} v_n$. The label $\triangleright$ of this fact is determined by iteratively merging the labels along the path according to the rules in Table 1.

---

[3] Interpreted function symbols and constants are considered logical symbols.

Facts over the shared vocabulary V_S are the basic building blocks of interpolants for $\mathcal{L}$-graphs. In general, however, it is not possible to represent an interpolant for $F \wedge G$ as an $\mathcal{L}$-graph or as an $\mathcal{L}$-formula (see Remark 1 in Section 2). Intuitively, the reason is that the proof of inconsistency is a result of a mutual interplay[4] of facts derived from F-coloured as well as from G-coloured edges. An F-coloured congruence edge may be derived from a path that contains edges corresponding to atoms in G. This prevents us from extracting a contradictory sub-graph of the proof that is derived exclusively from F.

We account for the interrelation between F-coloured and G-coloured facts by introducing conditions and premises for facts in $\mathcal{L}$-graphs.

Definition 4 (Conditions for facts, edges). *Let $E = E_F \uplus E_G$ and $V = V_F \cup V_G$ be a colouring of the edges and vertices of a proof of inconsistency for $F \wedge G$. A condition for a fact (or edge) $v_i \overset{=}{\rightarrow} v_j$ is a (possibly empty) set C of facts obtained from factorised and contracted paths in E such that one of the following conditions holds:*

- *$C = \emptyset$ and $v_i \overset{=}{\rightarrow} v_j$ is a contraction of edges corresponding to atoms in $F \wedge G$.*
- *$v_i \overset{=}{\rightarrow} v_j$ can be derived from the $\mathcal{L}$-graph $\mathcal{G}(V,C)$ by means of equality and congruence closure and equality-entailing cycles.*

We refer to the subset of F-coloured (G-coloured) facts in C as F-condition (G-condition, respectively).

The facts in a proof of inconsistency as constructed by the decision procedure in Section 3 comprise congruence edges, edges derived from equality-entailing cycles, and "basic" edges corresponding to atoms in the original formula $F \wedge G$. The conditions for basic edges and facts comprising only basic edges are defined to be $C = \emptyset$ in Def. 4. For the remaining artefacts, we construct a set of conditions C as follows:

1. *Congruence edges.* For a congruence edge, C is the set of facts obtained by factorising and contracting the path the congruence edge is derived from.
2. *Edges derived from equality-entailing cycles.* For an edge derived from an equality-entailing cycle, C is the set of facts obtained by factorising and contracting that cycle.
3. *Facts.* The condition for a fact $v_1 \rightarrow v_n$ obtained by contracting the path $v_1 \rightarrow \ldots \rightarrow v_n$ is $\bigcup_{i \in \{1..n-1\}} C_i$, where C_i is a condition for $v_i \rightarrow v_{i+1}$.

The correctness of this construction follows immediately from Def. 4 and the definition of congruence edges and derived edges.

A premise denotes a *recursively closed* set of conditions, in which the derived facts are in turn justified by their respective conditions:

---

[4] This process can also be formalised as a cooperative two-player game [4].

Definition 5 (Premises for facts). *The F-premise for a fact $v_i \overset{=}{\to} v_j$ of colour G is the set F-premise$(v_i \overset{=}{\to} v_j)$ of F-coloured facts defined as*

$$F\text{-}premise\,(v_i \overset{=}{\to} v_j) \overset{\text{def}}{=}$$

$$(F\text{-}condition\ for\ v_i \overset{=}{\to} v_j)\ \cup$$

$$\bigcup\{F\text{-}premise\,(v_n \to v_m)\mid v_n \to v_m \in (G\text{-}condition\ for\ v_i \overset{=}{\to} v_j)\}.$$

The definition of the G-premise for F-coloured facts is symmetric.

Premises can be seen as a form of rely-guarantee reasoning. F-premises take the role of ρ in McMillan's interpolations [3], and G-premises correspond to justifications in [4].

Lemma 2. *Let C be the F-condition of a G-coloured fact (or edge) $v_i \overset{=}{\to} v_j$ in a coloured proof of inconsistency $\mathcal{G}(E, V)$, where $E = E_F \cup E_G$ and $V = V_F \cup V_G$. For all $(v_n \to v_m) \in C$ it holds that $v_n, v_m \in V_S$.*

It follows immediately that F-premises and G-premises refer only to the shared vertices of a proof of inconsistency (cf. Lemma 2(iii) in [4]).

Definition 6 ($\mathcal{L}$-graph-based interpolant). *Let $\mathcal{G}(V, E)$ be a proof of inconsistency for $F \wedge G$ and let $E = E_F \cup E_G$ and $V = V_F \cup V_G$, $V_S = V_F \cap V_G$ be a colouring of its edges and vertices. A $\mathcal{L}$-graph-based interpolant is a pair $\langle \mathcal{I}, \mathcal{J}\rangle$ of sets such that the following mutual conditions hold:*

1. *$\mathcal{J}$ is a set of pairs $\langle P, v_i \to v_j\rangle$, and for each $\langle P, v_i \to v_j\rangle \in \mathcal{J}$ it holds that*
   (a) *$P \subseteq \mathcal{I}$ is the F-premise for the G-coloured fact $v_i \to v_j$, and*
   (b) *for all $v_n \to v_m \in \mathcal{I}$, the G-premise for $v_n \to v_m$ is a subset of*

   $$\{v_k \to v_l \mid \langle P, v_k \to v_l\rangle \in \mathcal{J}\}.$$

2. *$\mathcal{I}$ is a set of F-coloured facts obtained by contracting edges in E_F, and the graph*

   $$\mathcal{G}\,(V_S, \mathcal{I} \cup \{v_i \to v_j \mid \langle P, v_i \to v_j\rangle \in \mathcal{J}\}) \tag{1}$$

   contains a contradictory cycle.
3. *For all $v_n \overset{\rhd}{\to} v_m$ in $\mathcal{I} \cup \{v_i \overset{\rhd}{\to} v_j \mid \langle P, v_i \overset{\rhd}{\to} v_j\rangle \in \mathcal{J}\}$ it holds that either*
   (a) *$v_n, v_m \in V_S$, or*
   (b) *$v_n \simeq v_m$ and $\rhd \in \{>, \neq\}$.*

Fig. 7 shows an algorithm that extracts a pair $\langle \mathcal{I}, \mathcal{J}\rangle$ from a proof of inconsistency. We argue that $\langle \mathcal{I}, \mathcal{J}\rangle$ is an $\mathcal{L}$-graph-based interpolant:

1. Since the factorisation and contraction preserves the structure of the graph, the graph $\mathcal{G}(V_F \cup V_G, E_F \cup E_G)$ contains a contradictory cycle of facts (possibly degenerate, i.e., $v_i \overset{\rhd}{\to} v_i$, $\rhd \in \{>, \neq\}$). Therefore, E_C exists.

1: **let** $\mathcal{G}(V_F \cap V_G, E_F \uplus E_G)$ **be** the factorised and contracted proof
2: **let** E_C **be** the facts in the contradictory cycle of $\mathcal{G}(V_F \cup V_G, E_F \uplus E_G)$
3: $\mathcal{W} := E_C$, $\mathcal{I} := \emptyset$, $\mathcal{J} := \emptyset$
4: **while** $(\mathcal{W} \neq \emptyset)$ **do**
5:      remove $v_i \rightarrow v_j$ from $\mathcal{W}$
6:      **if** $v_i \rightarrow v_j$ is G-coloured **then**
7:           $P := F\text{-premise}\,(v_i \rightarrow v_j)$
8:           $\mathcal{J} := \mathcal{J} \cup \{\langle P, v_i \rightarrow v_j \rangle\}$
9:      **else**
10:           $P := G\text{-premise}\,(v_i \rightarrow v_j)$
11:           $\mathcal{I} := \mathcal{I} \cup \{v_i \rightarrow v_j\}$
12:      **end if**
13:      $\mathcal{W} := \mathcal{W} \cup P$
14: **end while**

Fig. 7. Computing an $\mathcal{L}$-graph-based interpolant

2. Observe that the algorithm maintains the following invariants:

   (a) For each $\langle P, v_i \rightarrow v_j \rangle \in \mathcal{J}$, P is an F-premise of $v_i \rightarrow v_j$ and a subset of $\mathcal{W} \cup \mathcal{I}$ (established in line 3 and maintained by lines 7, 8, and 13).
   (b) For each $v_i \rightarrow v_j \in \mathcal{I}$, the G-premise of $v_i \rightarrow v_j \in \mathcal{I}$ is a subset of $\mathcal{W} \cup \{v_n \rightarrow v_m \mid \langle P, v_n \rightarrow v_m \rangle \in \mathcal{J}\}$. This is established in line 3 and maintained by the statements in lines 10, 11, and 13.
   (c) $\mathcal{G}(V_S, \mathcal{W} \cup \mathcal{I} \cup \{v_i \rightarrow v_j \mid \langle P, v_i \rightarrow v_j \rangle \in \mathcal{J}\})$ contains a contradictory cycle. This invariant is established in line 3.

   Upon termination of the algorithm, $\mathcal{W} = \emptyset$ holds. Together with $\mathcal{W} = \emptyset$, the invariant (2a) implies condition (1a) and the invariant (2b) implies condition (1b) in Def. 6. Furthermore, it follows from the invariant (2c) that condition (2) in Def. 6 is fulfilled.
3. Due to the tree-structured derivations in the proof, the algorithm terminates.
4. The sets $\mathcal{I}$ and $\mathcal{J}$ contain only

   (a) edges $v_i \rightarrow v_j$ from the factorised and contracted contradictory cycle E_C of the proof of inconsistency (lines 2 and 3), and
   (b) G-premises (F-premises) for F-coloured (G-coloured) facts.

   According to Lemma 2, all facts in the premises (4b) are edges with endpoints $v_i, v_j \in V_S$. If E_C contains F-coloured as well as G-coloured facts, then the facts (4a) must be edges connecting vertices in V_S. Otherwise, E_C contains a single degenerate edge $v_i \xrightarrow{\triangleright} v_i$, where v_i is not necessarily an element of V_S. Therefore, condition 3 in Def. 6 holds.

The interpolant I for an $\mathcal{L}$-formula $F \wedge G$ may not be expressible in $\mathcal{L}$ (see Remark 1). We can, however, translate the $\mathcal{L}$-graph-based interpolant into an $\mathcal{L}$-formula with disjunctions:

(a) A proof of inconsistency

$$f(g(y)) = v \wedge w = y \vee$$
$$f(g(y)) = v \vee$$
$$(x \neq u) \vee$$
$$(w = y) \wedge g(z) \neq g(y)$$

(b) Interpolant for Fig. 8a

Fig. 8. An example of an interpolant for an inconsistency proof

$$I \stackrel{\text{def}}{=} \underbrace{\bigwedge_{v_i \stackrel{\triangleright}{\to} v_j \in \mathcal{I}} (t_i \triangleright t_j)}_{(a)} \vee \underbrace{\bigvee_{\langle P, v_n \stackrel{\triangleright}{\to} v_m \rangle \in \mathcal{J}} \left(\bigwedge_{(v_i \stackrel{\triangleright_P}{\to} v_j) \in P} (t_i \triangleright_P t_j) \right) \wedge \neg(t_n \triangleright t_m)}_{(b)} \quad (2)$$

We simplify all terms of the form $t_i \triangleright t_i$ to **false** if $\triangleright \in \{>, \neq\}$ and to **true** if $\triangleright \in \{\geq, =\}$.

Example 2. Consider the proof of inconsistency shown in in Fig. 8a. Contracting the inconsistent cycle yields $f(g(y)) \neq v$ (G-coloured) and $f(g(y)) = v$ (F-coloured) *under the condition* that $u = g(z)$ (F-coloured), and $g(z) = g(y)$ (G-coloured) hold. The condition for $g(z) = g(y)$, in turn, is that $z = w$ and $w = y$ holds, where the latter fact is F-coloured. The resulting interpolant is shown in Fig. 8b.   ◁

Finally, we claim that I as defined in (2) is indeed an interpolant for $F \wedge G$.

Theorem 1. *Given an $\mathcal{L}$-graph-based interpolant $\langle \mathcal{I}, \mathcal{J} \rangle$ for an $\mathcal{L}$-formula $F \wedge G$, the formula (2) is an interpolant for $F \wedge G$.*

Let us provide an intuitive explanation of Formula (2) before we proceed to the proof of Theorem 1. The formula is split into two sub-formulæ (a) and (b): Condition (1b) in Def. 6 guarantees that $(2a)$ holds if

$$\bigwedge_{\langle P, v_n \stackrel{\triangleright}{\to} v_m \rangle \in \mathcal{J}} (t_n \triangleright t_m) \quad (3)$$

and F (i.e., the F-coloured atoms in $F \wedge G$) hold.

Formula (2b) takes the rôle of the interface in rely-guarantee reasoning and challenges G to contradict one of the atoms in Formula (3). The F-premises of these G-coloured atoms are a subset of $\mathcal{I}$, and therefore implied by $(2a)$ due to condition (1a) in Def. 6. The G-premises of the facts in $\mathcal{I}$ are in turn implied by Formula (3). The tree-structured derivations of congruence edges and derived edges (generated by algorithm in Fig. 7) prevent circular reasoning. The resulting tree-structure of these premises is illustrated in Fig. 9: The G-coloured facts $e_1, \ldots, e_5$ derived from the F-premises at the leaves in turn form the G-premise at an inner node of the tree, and so on. We show that this structure prevents G from contradicting Formula (3).

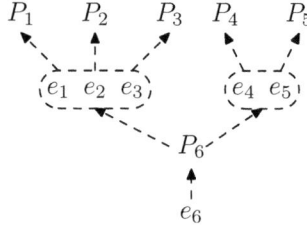

Fig. 9. Tree-structure of the premises in an interpolant

We are now in a position to show the correctness of Theorem 1.

Proof: We review the conditions of Def. 1 in Section 2:

1. $F \models I$. Consider that the G-premises (3) for the F-facts in (2a) hold. W.l.o.g., pick an edge $v_i \overset{\triangleright}{\to} v_j$ from $\mathcal{I}$. Since (3) holds, the G-premise for $v_i \overset{\triangleright}{\to} v_j$ holds. We show that the F-condition (Def. 4) for $v_i \overset{\triangleright}{\to} v_j$ is implied by F and (3) by means of induction.

   - *Base case.* The height of the tree-shaped derivation of $v_i \overset{\triangleright}{\to} v_j$ is one; Thus, the F-condition of $v_i \overset{\triangleright}{\to} v_j$ is a subset of the atoms in F.
   - *Hypothesis.* The F-condition of $v_i \overset{\triangleright}{\to} v_j$ is implied by F and (3) if the height of the tree-shaped derivation is $n - 1$ or less.
   - *Induction step.* The height of the derivation of $v_i \overset{\triangleright}{\to} v_j$ is n. W.l.o.g., pick a fact $v_n \to v_m$ from the F-condition of $v_i \overset{\triangleright}{\to} v_j$. The height of the tree-shaped derivation of this fact is $n - 1$ or less. The G-condition for $v_n \to v_m$ holds because of Def. 5, condition (1b) in Def. 6, and the assumption that (3) holds. The F-condition of $v_n \to v_m$ holds by our induction hypothesis, and therefore $v_n \to v_m$ and the F-condition of $v_i \overset{\triangleright}{\to} v_j$ must hold.

   Therefore, F and (3) imply (2a). Otherwise, at least one atom $t_i \triangleright t_j$ in (3) is false. W.l.o.g., we can choose a fact $v_i \overset{\triangleright}{\to} v_j$ (e.g., e_6 in Fig. 9) such that the following conditions hold:

   - $v_i \overset{\triangleright}{\to} v_j$ corresponds to an atom $t_i \triangleright t_j$ in (3) which is false.
   - The G-premises of the F-premise of $v_i \overset{\triangleright}{\to} v_j$ comprise only of facts corresponding to atoms in (3) that are true.

   Then, using the same induction argument as above, we can show that the F-premise P for the G-coloured fact $v_i \overset{\triangleright}{\to} v_j$ holds. Therefore, the conjunct corresponding to $\langle P, v_i \overset{\triangleright}{\to} v_j \rangle$ is true.

2. $G \wedge I \models \bot$. Assume that the formulæ $(2a)$ and (3) hold, i.e., G does not contradict (3). Since $(2a)$ corresponds to $\mathcal{I}$ and (3) to

$$\{v_i \to v_j \,|\, \langle C, v_i \to v_j \rangle \in \mathcal{J}\},$$

$G \wedge I$ must be contradictory (condition 2 in Def. 6).

Otherwise, in order for G to contradict (3), at least one of the atoms in $\{t_n \rhd t_m \,|\, \langle P, v_n \overset{\rhd}{\to} v_m \rangle \in \mathcal{J}\}$ must be false. Using induction, we show that the condition of $v_n \overset{\rhd}{\to} v_m$ holds, contradicting the assumption that $\neg(t_n \rhd t_m)$ holds. Thanks to condition $(1b)$ in Def. 6, the F-premise of $v_n \overset{\rhd}{\to} v_m$ is a subset of $\mathcal{I}$. It remains to show that the G-condition of $v_n \overset{\rhd}{\to} v_m$ holds.

 - *Base case.* The height of the tree-shaped derivation of $v_i \overset{\rhd}{\to} v_j$ is one; Thus the G-condition of $v_i \overset{\rhd}{\to} v_j$ is a subset of the atoms in G.
 - *Hypothesis.* The G-condition of $v_i \overset{\rhd}{\to} v_j$ is implied by G and Formula $(2a)$ if the height of the tree-shaped derivation is $n - 1$ or less.
 - *Induction step.* The height of the derivation of $v_i \overset{\rhd}{\to} v_j$ is n. W.l.o.g., pick a fact $v_n \to v_m$ from the G-condition of $v_i \overset{\rhd}{\to} v_j$. The height of the tree-shaped derivation of this fact is $n - 1$ or less. The F-condition for $v_n \to v_m$ holds because of Def. 5, condition $(1a)$ in Def. 6, and the assumption that Formula $(2a)$ holds. The G-condition of $v_n \to v_m$ holds by our induction hypothesis, and therefore $v_n \to v_m$ and the G-condition of $v_i \overset{\rhd}{\to} v_j$ must hold.

3. Condition 3 in Def. 6 and the fact that we simplify terms $t_i \rhd t_i$ to **true** or **false** guarantee that I refers only to shared variables and function symbols. ∎

The next section discusses applications of our interpolating decision procedure and provides an evaluation of its adequacy for verifying systems software.

5   Application and Evaluation

The two most prominent interpolation-based software model checking techniques are predicate abstraction [11] and interpolation-based abstraction [6]. Both techniques construct an abstract reachability tree by unwinding the (abstract) transition relation. The nodes in this tree are labelled with interpolants derived from infeasible counterexamples (i.e., unsatisfiable conjunctions of relations), thus over-approximating the set of safely reachable program states. The verification process terminates if a fixed-point of this set is reached.

The transition function of programs is typically represented using first order logic formulæ. The primitive data-types of a vast majority of programming languages have bounded domains. In order to be able to apply interpolation-based techniques in a *sound* manner, the decision procedure must not conclude that a formula is unsatisfiable if it is satisfiable in its bit-vector interpretation. This is not guaranteed if we use linear arithmetic over $\mathbb{R}$ or $\mathbb{Z}$: The operator $+$ in infinite interpretations is addition on an infinite set, while it corresponds to addition mod m (for some m) in the case of bit-vectors. Consider the formula $a > b + 2 \wedge a \leq b$ over the 2-bit variables a and b. This formula has the satisfying assignment $\{a \mapsto 2, b \mapsto 2\}$ in its bit-vector interpretation, while it is unsatisfiable in the theory of linear arithmetic over the reals or the integers.

While the ability of our algorithm to handle arithmetic operations is very limited (our rewriting rules can simplify terms involving addition in only certain special cases), it does not falsely conclude unsatisfiablity for the bit-vector interpretation. However, we may fail to prove unsatisfiability in certain cases (for instance, a chain of 2^n disequalities over $2^n + 1$ distinct n-bit variables). The reason underlying this problem is that the Nelson-Oppen method requires theories to be stably infinite, which is not the case for the theory of bit-vectors. This may lead to spurious counterexamples, which can be caught by falling back to a bit-level accurate decision procedure (such as bit-flattening [12]).

Finally, we have to ask whether our logic is *sufficient* to represent the transition relation of realistic programs. Whether the relations and interpreted functions provided by $\mathcal{L}$ are sufficient depends largely on the application domain. A common benchmark for software model checking tools is the set of Windows device drivers used in [6]. In order to evaluate the usefulness of our logic $\mathcal{L}$, we have integrated the decision procedure into our prototypical interpolation-based model checker WOLVERINE. WOLVERINE is an implementation of the algorithm presented in [6]. It generates conjunctive formulæ by unwinding the program and labelling the edges in the reachability graph with transition relations. In this setting, formulæ corresponding to infeasible paths are unsatisfiable. We ran WOLVERINE on the `kbfiltr.i`, `floppy.i`, and `mouclass.i` drivers presented in [13,6]. Our decision procedure was able to provide interpolants for all unsatisfiable formulæ encountered during the verification process.[5] We attribute this to the fact that device drivers make little use of arithmetic. The loops typically iterate over initialised induction variables, which can be handled by constant propagation (resulting in ground terms that can be rewritten).

6 Related Work

The related work in the area of decision procedures is vast. We focus on recent *interpolating* decision procedures. The first implementation of an interpolating decision procedure widely used in verification is McMillan's FOCI [3]. This tool supports linear arithmetic over $\mathbb{R}$ and equality with uninterpreted functions (EUF), and introduces the semantic discrepancy discussed in Section 5 when used for program verification. Based on the ideas in [3], Fuchs presents a graph-based approach for EUF [4]. The interpolants in CNF generated by this technique are reported to be (syntactically) smaller than the results of FOCI. In comparison, we support a strict super-set of EUF and generate interpolants in DNF. Fuchs' work has recently been extended to combined theories [5], and our algorithm can be seen as an instance of that framework. An interpolating

---

[5] We do not present results on the run-time, as the performance of WOLVERINE is not yet competitive due to a lack of optimisation of the model checking algorithm.

decision procedure for the theory of unit-to-variable-per-inequality ($\mathcal{UTVPI\star}$), a logic with atoms of the form $(0 \leq ax_1 + bx_2 + k)$ over $\mathbb{Z}$, is presented in [14]. Jain et al. present an interpolating decision procedure for linear modular equations [15], but does not support uninterpreted functions. We plan to integrate this algorithm into our implementation.

Our algorithm can also be implemented in a Nelson-Oppen or SMT framework, and interpolants can be generated using the mechanisms presented in [10] or [5,16]. It can also be integrated in a *proof-lifting* decision procedure, which constructs word-level proofs from propositional resolution proofs [12].

7   Conclusion and Future Work

We present a decision procedure for a first-order logic fragment with the relations $=$, $\neq$, $\geq$, and $>$ and argue that this logic is an efficiently decidable subset of first order logic. Furthermore, the logic is sound with respect to reasoning about software with bounded integers. We intend to perform an evaluation of a larger scale than presented in this paper. Furthermore, we plan to integrate acceleration techniques similar to [17] into our interpolation-based model checker WOLVERINE.

Acknowledgements. We thank Philipp Rümmer and May Chan for their detailed comments on our paper and Mitra Purandare for the inspiring discussions.

References

1. McMillan, K.L.: Applications of Craig interpolation to model checking. In: Marcinkowski, J., Tarlecki, A. (eds.) CSL 2004. LNCS, vol. 3210, pp. 22–23. Springer, Heidelberg (2004)
2. Pudlák, P.: Lower bounds for resolution and cutting plane proofs and monotone computations. The Journal of Symbolic Logic 62, 981–998 (1997)
3. McMillan, K.L.: An interpolating theorem prover. Theoretical Computer Science 345, 101–121 (2005)
4. Fuchs, A., Goel, A., Grundy, J., Krstić, S., Tinelli, C.: Ground interpolation for the theory of equality. In: Kowalewski, S., Philippou, A. (eds.) TACAS 2009. LNCS, vol. 5505, pp. 413–427. Springer, Heidelberg (2009)
5. Goel, A., Krstić, S., Tinelli, C.: Ground interpolation for combined theories. In: Schmidt, R.A. (ed.) CADE-22. LNCS, vol. 5663, pp. 183–198. Springer, Heidelberg (2009)
6. McMillan, K.L.: Lazy abstraction with interpolants. In: Ball, T., Jones, R.B. (eds.) CAV 2006. LNCS, vol. 4144, pp. 123–136. Springer, Heidelberg (2006)
7. Cimatti, A., Sebastiani, R.: Building efficient decision procedures on top of SAT solvers. In: Bernardo, M., Cimatti, A. (eds.) SFM 2006. LNCS, vol. 3965, pp. 144–175. Springer, Heidelberg (2006)
8. Meir, O., Strichman, O.: Yet another decision procedure for equality logic. In: Etessami, K., Rajamani, S.K. (eds.) CAV 2005. LNCS, vol. 3576, pp. 307–320. Springer, Heidelberg (2005)
9. Nieuwenhuis, R., Oliveras, A.: Proof-Producing Congruence Closure. In: Giesl, J. (ed.) RTA 2005. LNCS, vol. 3467, pp. 453–468. Springer, Heidelberg (2005)

10. Yorsh, G., Musuvathi, M.: A combination method for generating interpolants. In: Nieuwenhuis, R. (ed.) CADE 2005. LNCS (LNAI), vol. 3632, pp. 353–368. Springer, Heidelberg (2005)
11. Henzinger, T.A., Jhala, R., Majumdar, R., McMillan, K.L.: Abstractions from proofs. In: Principles of Programming Languages, pp. 232–244. ACM, New York (2004)
12. Kroening, D., Weissenbacher, G.: Lifting Propositional Interpolants to the Word-Level. In: Formal Methods in Computer-Aided Design, pp. 85–89. IEEE, Los Alamitos (2007)
13. Henzinger, T.A., Jhala, R., Majumdar, R., Sutre, G.: Lazy abstraction. In: Principles of Programming Languages, pp. 58–70. ACM, New York (2002)
14. Cimatti, A., Griggio, A., Sebastiani, R.: Interpolant generation for UTVPI⋆. In: Schmidt, R.A. (ed.) CADE-22. LNCS, vol. 5663, pp. 167–182. Springer, Heidelberg (2009)
15. Jain, H., Clarke, E.M., Grumberg, O.: Efficient craig interpolation for linear diophantine (dis)equations and linear modular equations. In: Gupta, A., Malik, S. (eds.) CAV 2008. LNCS, vol. 5123, pp. 254–267. Springer, Heidelberg (2008)
16. Bruttomesso, R., Cimatti, A., Franzén, A., Griggio, A., Sebastiani, R.: Delayed theory combination vs. Nelson-Oppen for satisfiability modulo theories: A comparative analysis. In: Hermann, M., Voronkov, A. (eds.) LPAR 2006. LNCS (LNAI), vol. 4246, pp. 527–541. Springer, Heidelberg (2006)
17. Kroening, D., Weissenbacher, G.: Verification and falsification of programs with loops using predicate abstraction. Formal Aspects of Computing (2009); (published Online First^{TM})

Author Index